Occupational Health and Safety in the Care and Use of Research Animals

Committee on Occupational Safety and Health in
Research Animal Facilities

Institute of Laboratory Animal Resources

Commission on Life Sciences

National Research Council

NATIONAL ACADEMY PRESS
Washington, DC   1997

NATIONAL ACADEMY PRESS • 2101 Constitution Avenue, NW • Washington, DC 20418

NOTICE: The project that is the subject of this report was approved by the Governing Board of the National Research Council, whose members are drawn from the councils of the National Academy of Sciences, National Academy of Engineering, and Institute of Medicine. The members of the committee responsible for the report were chosen for their special competences and with regard for appropriate balance.

This report has been reviewed by a group other than the authors according to procedures approved by a Report Review Committee consisting of members of the National Academy of Sciences, National Academy of Engineering, and Institute of Medicine.

This study was supported under contract number NO1-RR-2-2118 by the National Center for Research Resources (NCRR), National Institutes of Health, Department of Health and Human Services (DHHS), which served as the lead agency for DHHS funding received from the Centers for Disease Control and Prevention, National Cancer Institute, National Institute on Aging, National Institute of Arthritis and Musculoskeletal and Skin Diseases, National Institute of Diabetes and Digestive and Kidney Diseases, National Institute of Environmental Health Sciences, National Institute of General Medical Sciences, National Institute of Neurological Disorders and Stroke, and the Office for Protection from Research Risks. NCRR also served as the lead agency for receipt of funding from the Agricultural Research Service, U.S. Department of Agriculture, and the Veterans Administration. Financial support was also provided by Research Resources, Merck Research Labs.

Core support is provided to the Institute of Laboratory Animal Resources by the Comparative Medicine Program, National Center for Research Resources, National Institutes of Health, through grant number 5P40RR0137; the National Science Foundation through grant number BIR-9024967; the U.S. Army Medical Research and Development Command, which serves as the lead agency for combined U.S. Department of Defense funding also received from the Human Systems Division of the U.S. Air Force Systems Command, Armed Forces Radiobiology Research Institute, Uniformed Services University of the Health Sciences, and U.S. Naval Medical Research and Development Command, through grant number DAMD17-93-J-3016; the American Cancer Society through grant number RC-1-34; and the American College of Laboratory Animal Medicine.

Any opinions, findings, and conclusions or recommendations expressed in this publication do not necessarily reflect the views of DHHS or other sponsors, nor does the mention of trade names, commercial products, or organizations imply endorsement by the U.S. government or other sponsors.

Occupational Health and Safety in the Care and use of Research Animals is available from the National Academy Press, 2101 Constitution Avenue, NW, Box 285, Washington DC 20055 1-800-624-6242; 202-334-3313 (in the Washington Metropolitan Area); http://www.nap.edu

Library of Congress Cataloging-in-Publication Data

Occupational health and safety in the care and use of research animals / Committee on
    Occupational Safety and Health in Research Animal Facilities, Institute of Laboratory Animal
    Resources, Commission on Life Sciences, National Research Council.
        p. cm.
    Includes bibliographical references and index.
    ISBN 0-309-05299-8 (pbk.)
        1. Laboratory animal technicians—Health risk assessment.
    2. Animal health technicians—Health risk assessment.
    3. Occupational diseases—Prevention.   I. Institute of Laboratory Animal Resources (U.S.).
    Committee on Occupational Safety and Health in Research Animal Facilities.
    RC965.A6O23 1997
    363.11'9619—dc21   97-4794

Copyright 1997 by the National Academy of Sciences. All right reserved.
Printed in the United States of America

COMMITTEE ON OCCUPATIONAL SAFETY AND HEALTH IN RESEARCH ANIMAL FACILITIES

W. Emmett Barkley (*Chair*), Laboratory Safety, Howard Hughes Medical Institute, Chevy Chase, Maryland
Rebecca Bascom, University of Maryland School of Medicine, Baltimore, Maryland
Robert K. Bush, Allergy Section, University of Wisconsin, and William S. Middleton VA Hospital, Madison, Wisconsin
Diane O. Fleming, The Johns Hopkins University, Baltimore, Maryland
Peter J. Gerone, Tulane Regional Primate Research Center, Tulane University Medical Center, Covington, Louisiana
Janet C. Gonder, Baxter Healthcare Corporation, Round Lake, Illinois
A. Wallace Hayes, The Gillette Company, Boston, Massachusetts
Julia K. Hilliard, Department of Virology and Immunology, Southwest Foundation for Biomedical Research, San Antonio, Texas
Christian E. Newcomer, Division of Laboratory Animal Medicine, School of Medicine, The University of North Carolina, Chapel Hill, North Carolina
James H. Stewart, Harvard University and Data Chem Software, Westboro, Massachusetts
Wayne R. Thomann, Department of Occupational and Environmental Safety, Duke University, Durham, North Carolina

Staff

Thomas L. Wolfle, Program Director
Ralph Dell, Visiting Scientist
Amanda Hull, Project Assistant (through 1995)
Cheryl Mitchell, Project Assistant
Carol M. Rozmiarek, Project Assistant
Norman Grossblatt, Editor

INSTITUTE OF LABORATORY ANIMAL RESOURCES COUNCIL

John L. VandeBerg (*Chair*), Southwest Foundation for Biomedical Research, San Antonio, Texas
Christian R. Abee, University of South Alabama, Mobile, Alabama
Muriel T. Davisson, The Jackson Laboratory, Bar Harbor, Maine
Bennett Dyke, Southwest Foundation for Biomedical Research, San Antonio, Texas
Neal L. First, University of Wisconsin, Madison, Wisconsin
Gerald F. Gebhart, University of Iowa, Iowa City, Iowa
James W. Glosser, Massillon, Ohio
John P. Hearn, Wisconsin Regional Primate Research Center, Madison, Wisconsin
Margaret S. Landi, SmithKline Beecham Pharmaceuticals, King of Prussia, Pennsylvania
Charles R. McCarthy, Kennedy Institute of Ethics, Georgetown University, Washington, DC
Robert J. Russell, Harlan Sprague Dawley, Indianapolis, Indiana
Richard C. Van Sluyters, University of California, Berkeley, California
John G. Vanderbergh, North Carolina State University, Raleigh, North Carolina
Peter A. Ward, University of Michigan Medical School, Ann Arbor, Michigan
Thomas D. Pollard, The Salk Institute for Biological Studies, La Jolla, California (ex officio member)

Staff

Thomas L. Wolfle, Program Director
Mara L. Glenshaw, Research Assistant
Cheryl Mitchell, Project Assistant
Carol M. Rozmiarek, Project Assistant

The Institute of Laboratory Animal Resources (ILAR) was founded in 1952 under the auspices of the National Research Council. A component of the Commission on Life Sciences, ILAR develops guidelines and disseminates information on the scientific, technological, and ethical use of animals and related biological resources in research, testing, and education. ILAR promotes high-quality, humane care of animals and the appropriate use of animals and alternatives. ILAR functions within the mission of the National Academy of Sciences as an adviser to the federal government, the biomedical research community, and the public.

COMMISSION ON LIFE SCIENCES

Thomas D. Pollard (*Chairman*), The Salk Institute for Biological Studies, La Jolla, California
Frederick R. Anderson, Cadwalader, Wickersham & Taft, Washington, D.C.
John C. Bailar III, University of Chicago, Chicago, Illinois
Paul Berg, Stanford University School of Medicine, Stanford, California
John E. Burris, Marine Biological Laboratory, Woods Hole, Massachusetts
Sharon L. Dunwoody University of Wisconsin, Madison, Wisconsin
Ursula W. Goodenough, Washington University, St. Louis, Missouri
Henry W. Heikkinen, University of Northern Colorado, Greeley, Colorado
Hans J. Kende, Michigan State University, East Lansing, Michigan
Susan E. Leeman, Boston University School of Medicine, Boston, Massachusetts
Thomas E. Lovejoy, Smithsonian Institution, Washington, D.C.
Donald R. Mattison, University of Pittsburgh, Pittsburgh, Pennsylvania
Joseph E. Murray, Wellesley Hills, Massachusetts
Edward E. Penhoet, Chiron Corporation, Emeryville, California
Emil A. Pfitzer, Research Institute for Fragrance Materials, Inc., Hackensack, New Jersey
Malcolm C. Pike, University of Southern California Comprehensive Cancer Center, Los Angeles, California
Henry C. Pitot III, McArdle Laboratory for Cancer Research, Madison, Wisconsin
Jonathan M. Samet, The Johns Hopkins University, Baltimore, Maryland
Charles Stevens, The Salk Institute for Biological Studies, La Jolla, California
John L. VandeBerg, Southwest Foundation for Biomedical Research, San Antonio, Texas

Staff: **Paul Gilman**, Executive Director

The National Academy of Sciences is a private, nonprofit, self-perpetuating society of distinguished scholars engaged in scientific and engineering research, dedicated to the furtherance of science and technology and to their use for the general welfare. Upon the authority of the charter granted to it by the Congress in 1863, the Academy has a mandate that requires it to advise the federal government on scientific and technical matters. Dr. Bruce Alberts is president of the National Academy of Sciences.

The National Academy of Engineering was established in 1964, under the charter of the National Academy of Sciences, as a parallel organization of outstanding engineers. It is autonomous in its administration and in the selection of its members, sharing with the National Academy of Sciences the responsibility for advising the federal government. The National Academy of Engineering also sponsors engineering programs aimed at meeting national needs, encourages education and research, and recognizes the superior achievements of engineers. Dr. William A. Wulf is president of the National Academy of Engineering.

The Institute of Medicine was established in 1970 by the National Academy of Sciences to secure the services of eminent members of appropriate professions in the examination of policy matters pertaining to the health of the public. The Institute acts under the responsibility given to the National Academy of Sciences by its congressional charter to be an adviser to the federal government and upon its own initiative to identify issues of medical care, research, and education. Dr. Kenneth I. Shine is president of the Institute of Medicine.

The National Research Council was established by the National Academy of Sciences in 1916 to associate the broad community of science and technology with the Academy's purposes of furthering knowledge and advising the federal government. Functioning in accordance with general policies determined by the Academy, the Council has become the principal operating agency of both the National Academy of Sciences and National Academy of Engineering in the conduct of their services to the government, the public, and the scientific and engineering communities. The Council is administered jointly by both Academies and the Institute of Medicine. Dr. Bruce Alberts and Dr. William A. Wulf are chairman and vice-chairman, respectively, of the National Research Council.

Preface

Occupational health and safety has long been a priority in the nation's research enterprise and of the National Research Council (NRC) of the National Academy of Engineering and the National Academy of Sciences. Over the last 2 decades, the NRC has provided substantive guidance in environmental health and safety to laboratory workers, managers, and government policy-makers through four major reports: *Prudent Practices for Handling Hazardous Chemicals in Laboratories* (1981), *Prudent Practices for Disposal of Chemicals from Laboratories* (1983), *Biosafety in the Laboratory: Prudent Practices for the Handling and Disposal of Infectious Materials* (1989), and *Prudent Practices in the Laboratory: Handling and Disposing of Chemicals* (1995) which consolidated and extensively revised the 1981 and 1983 reports. This tradition has now been extended to address occupational health and safety issues associated with the care and use of laboratory animals.

The Interagency Research Animal Committee (IRAC), composed of representatives of federal agencies that use or regulate the use of animals in research, asked the NRC to conduct a study and produce a report that would provide guidance for protecting the health and safety of workers who care for and use research animals. The need for such guidance was based both on the recognition of the broad array of occupational hazards in the specialized workplace of the animal research facility and on the absence of authoritative guidance that institutions could use to develop appropriate occupational health and safety programs within their animal research facilities. The IRAC and NRC considered this study particularly important because grantees of the US Public Health Service are required to address the need for an occupational health program as recommended in the *Guide for the Care and Use of Laboratory Animals* and particularly timely because the *Guide* was scheduled for revision. The NRC appointed the Committee on Occupational Safety and Health of Personnel in Research Animal Facilities in January 1993. The study was conducted under the auspices of the Institute of Laboratory Animal Resources (ILAR) of the Commission on Life Sciences.

The committee was charged to provide guidelines for the development of occupational health and safety programs that would be suitable for all institutions that use research animals. Specific recommendations were requested of the committee on several relevant issues, including the need for periodic physical examinations, the value of serum banking, and who should be included in the animal research institution's occupational health and safety program.

This report differs considerably from its predecessors. Although it affirms prudent practices developed in the previous studies, the committee's approach has been to address the way in which prudent practices can best be incorporated into the animal care and use programs of research institutions. When hazards associated with laboratory research are viewed in the context of the animal facility, different strategies might be appropriate for achieving a safe and healthful workplace. A new set of workers, who might be less informed of research hazards, could become exposed to potentially hazardous experimental agents under circumstances quite different from the laboratory. The safety knowledge and expertise of the responsible laboratory worker might not be easily transferable to this new setting. And the use of research animals introduces new occupational health concerns, such as the risks of zoonoses and allergies to animals.

In the course of preparing this report, the committee met with a large number of specialists as an important part of its data-gathering. The committee hosted workshops in Washington, DC, and Irvine, California, with occupational health professionals; participated in the Forum on Occupational Health and Safety sponsored by the American College of Laboratory Animal Medicine; and conducted seminars at meetings of the American Association for Laboratory Animal Science, Public Responsibility in Medicine and Research (PRIM&R), and the Applied Research Ethics National Association (ARENA). Many people participated in those sessions and contributed substantially to the formulation of the committee's recommendations. To each of them the committee is greatly indebted. Special recognition is in order for the important and continuing assistance provided by Ralph Dell, Columbia University; Alan Ducatman, West Virginia University School of Medicine; Tom Ferguson, University of California, Davis; Suzi Goldmacher, University of California, San Francisco; George Jackson, Duke University; Thomas McBride, US Department of Energy; Albert E. New, Association for Assessment and Accreditation of Laboratory Animal Care, International; Jonathan Richmond and Margaret Tipple, Centers for Disease Control and Prevention; James Schmitt, National Institutes of Health; and Ellison Wittles, Baylor University College of Medicine.

Many letters of interest and support were received from people who struggle

PREFACE   ix

with occupational health issues at their institutions. They reminded us not to forget the small institutions, not to create costly bureaucracy, to help with the meaning of "substantial animal contact" in defining those who should be included in institutional occupational health and safety programs, and to help in determining needs for serum banking and other important parts of their occupational health programs. Their letters constituted a tremendous incentive to the committee and a constant reminder of the array of problems for which the report would be consulted.

We also want to acknowledge the contributions of the many individuals who willingly agreed to review our work. Their burden was our benefit as they thoughtfully improved the quality of this report.

The committee recognizes that this report will likely be revised in the future. It has been our intent to provide basic concepts and a valid foundation from which many models of successful occupational health and safety programs will emerge. Future revisions will benefit from this acquired experience. So we encourage readers who have evidence to support improved procedures or recommendations or who detect errors of omission or commission in this report to send their suggestions to the Institute of Laboratory Animal Resources, National Research Council, National Academy of Sciences, 2101 Constitution Avenue, NW, Washington, DC 20418.

The committee extends its appreciation to the sponsors of this report; to Norman Grossblatt for editing the manuscript; to James Glosser for his encouragement and wise counsel; to Carol Rozmiarek for her skillful support at each of the committee's meetings and for coordinating the great flow of information to and from committee members; to Amanda Hull for her steadfast assistance and polite reminders of our self-inflicted deadlines; and to Thomas Wolfle for his thoughtful nurturing, extraordinary tolerance, hard work, and firm belief that our good intentions would ultimately prevail.

                                         Emmett Barkley, *Chair*
                                         Committee on Occupational Safety and
                                         Health in Research Animal Facilities

Contents

1 INTRODUCTION, OVERVIEW, AND RECOMMENDATIONS   1
Introduction, 1
Overview, 3
Recommendations, 7

2 PROGRAM DESIGN AND MANAGEMENT   11
Program Goal, 11
Diversity, 11
Basic Concepts, 13
Accountability and Responsibility, 15
Institutional Activities and Their Interactions, 18
Management Style and Structure, 23
Getting Started, 23

3 PHYSICAL, CHEMICAL, AND PROTOCOL-RELATED HAZARDS   32
Physical Hazards, 32
Hazards Associated with Experimental Protocols, 43

4 ALLERGENS   51
Mechanisms of Allergic Reactions, 53
Specific Animals That Can Provoke Allergic Reactions, 54
Preventive Measures and Interventions, 60
Evaluation of the Allergic Worker, 63
Anaphylaxis, 64

5   ZOONOSES   65
    Viral Diseases, 66
    Rickettsial Diseases, 81
    Bacterial Diseases, 85
    Protozoal Diseases, 95
    Fungal Diseases, 99
    Helminth Infections, 101
    Arthropod Infestations, 101

6   PRINCIPAL ELEMENTS OF AN OCCUPATIONAL
    HEALTH AND SAFETY PROGRAM   106
    Administrative Procedures, 107
    Facility Design and Operation, 107
    Exposure Control Methods, 108
    Education and Training, 114
    Equipment Performance, 116
    Information Management, 118
    Emergency Procedures, 120
    Program Evaluation, 121

7   OCCUPATIONAL HEALTH-CARE SERVICES   123
    Federal Requirements and Guidelines for
       Occupational Health-Care Services, 124
    Assessment of Health Risks, 125
    Responsibilities of an Occupational Health-Care Service, 125
    Activities of an Occupational Health-Care Service, 129
    Program Evaluation, 133

REFERENCES   135

INDEX   147

1

Introduction, Overview, and Recommendations

INTRODUCTION

Much has been written about the obligation of institutions that care for and use research animals to be protective of the health and well-being of the animals. Thoughtful consideration of that institutional obligation has resulted in substantial improvements over the last 3 decades in animal-husbandry practices, facility design criteria, caging specifications, and institutional policies that govern the use of animals in research. But it has yielded little authoritative guidance for addressing a related institutional obligation—the protection of the health and safety of employees who care for and use research animals. Fortunately, progress in enhancing the quality of animal-care programs has had a beneficial effect in minimizing occupational-health risks of institutional employees.

This book by the Committee on Occupational Safety and Health in Research-Animal Facilities, in the Institute of Laboratory Animal Resources of the National Research Council's Commission on Life Sciences, is about the occupational health and safety of institutional employees, visitors, and students who in the course of their work with research animals might be exposed to hazards that could adversely affect their health and safety. Our task is to promote occupational health and safety by recognizing and considering hazards and health risks associated with the care and use of research animals. The book is written to be of assistance to institutions that are in the process of developing or re-evaluating occupational health and safety programs for employees engaged in animal care and use. The general concepts set forth apply to many categories of institutions: academic, industrial, and government research institutions; biomedical and agricultural research institutions; and medical and veterinary educational institutions. The book should also be useful to persons responsible for overseeing the health and safety of employees in related occupations, such as those in general veterinary practices, zoologic institutions, animal shelters, and kennels; employees in these categories face many of the same risks that are associated with the care and use of research animals.

Most research institutions have established environmental health and safety offices to foster institutional compliance with regulations promulgated by the Occupational Safety and Health Administration (OSHA) and the Environmental Protection Agency. Program emphasis has been placed on general awareness of hazardous chemicals, chemical safety in laboratories, control of bloodborne pathogens, and management of hazardous wastes. Collectively, this has been an overwhelming responsibility, with few resources available to develop the new program initiatives that are required to emphasize occupational health in animal care and use programs.

At the same time, the perceived need to provide occupational-health services, such as preassignment physical examinations and medical surveillance, to employees engaged in animal care and use has been driven by institutions' interpretations of broad regulatory requirements or contractual obligations imposed by funding agencies. Confusion as to what is needed and what is required has slowed the development of relevant program activities. Some institutions have undertaken expensive efforts to provide a broad array of occupational-health services, many of which have little benefit for the employee. Others have created health and safety programs that have no occupational-health component.

Artificial barriers to intra-institutional communication have often stifled the initiative of the persons in the institution who must become involved if worthy programs to promote occupational health and safety are to be developed. For example, the rigidity of jurisdictional boundaries has caused some managers of vivariums to avoid providing safety oversight and guidance to employees in other departments who conduct research in the vivariums. The institutional animal care and use committee (IACUC) is often the only forum for interaction, but its responsibilities are related only indirectly to occupational health and safety. Continuing collaboration among scientists, safety experts, health-care professionals, veterinarians, and administrators to promote health and safety has been difficult to initiate and difficult to sustain. Other barriers to effective program development involve several commonly unresolved questions: Who is responsible? Who will provide the necessary resources? Who has the authority to act?

We recognize that institutional management, particularly the direction and guidance provided by the senior official of an institution, is the key element required for developing and sustaining any useful occupational health and safety program. A truly successful program, however, will ultimately depend on the participation of all employees whose work might affect occupational health and safety—their own, their colleagues', their subordinates', or their co-workers'.

Thus, protecting the health and safety of employees engaged in the care and use of research animals is a cooperative enterprise that requires the active participation of institutional officials, scientists who plan and carry out research involving experimental animals, persons responsible for the management of animal care and use programs, health and safety professionals, and the individual employees themselves who must share the responsibility both for their own health and safety and for the health and safety of those around them. This volume is addressed to all who are responsible for the health and safety of employees engaged in the care and use of research animals.

We have tried not to be prescriptive in writing this book, and we do not present an ideal model for an occupational health and safety program. Such an approach would destroy our very premise: that there are many valid ways to fulfill an institution's commitment to provide a healthful and safe environment for an animal care and use workforce. The best model is one that accurately reflects the risks inherent in the research activities conducted at a specific institution and allows for the careful development and use of practical and relevant methods for controlling the hazards that contribute to those risks. The guidance provided here is intended to help individual institutions to address their own circumstances comprehensively and to determine their own best courses of action. This process demands good judgment and a genuine commitment to reduce risks to an acceptable level. It also requires objectivity because a careful consideration of local circumstances might result in the elimination of activities that were once but are no longer thought to be of value.

We expect institutions to choose to address the health and safety needs of animal care and use employees within the context of their existing environmental health and safety programs. But new initiatives must be the product of interactions among employees who represent administration, research, animal care and use, and occupational health. If not all these activities are involved in the development of program initiatives, the program will lose relevance, general acceptance, and effectiveness.

This book is designed to serve as an introductory guide to hazards associated with the care and use of research animals. We have tried to be comprehensive in our treatment of hazards that are inherently associated with the use of animals, such as allergens, zoonoses, and the obvious physical hazards, e.g., biting. We have dealt only briefly with hazard-control practices that are fully treated elsewhere. Readers will be guided to other references that should be consulted for further information on safe practices appropriate to their own circumstances.

OVERVIEW

The bulk of this book is divided into six chapters: Chapter 2 deals with how an institution addresses its responsibility for involving all employees and programs to meet the occupational health and safety concerns of all persons; Chapter

3 addresses physical, chemical, and protocol-related hazards associated with animal research; Chapter 4 provides a comprehensive discussion of allergic hazards; Chapter 5 summarizes relevant information on zoonoses of common research animals; Chapter 6 provides a general discussion of the key elements that are likely to be included in effective institutional programs; and Chapter 7 introduces concepts that are important in the development of the occupational-health element of the program. A brief overview of these chapters follows.

Chapter 2. Program Design and Management

The material in this chapter lays the foundation for developing an occupational health and safety program that addresses employee risks of illness and injury associated with the care and use of research animals. Program design requires an understanding of the tasks of at-risk employees; those employees' diversity in experience, education, and language proficiency; characteristics of the work environment; and the institutional mission. The work environment and mission are of paramount importance because they determine the nature of the hazards presented by the animal research activities.

The chapter defines the basic concepts that determine the effectiveness of an occupational health and safety program, which include the following:

- Knowing the hazard.
- Avoiding and controlling exposures.
- Training and education.
- Rules and guidelines.
- Consistency.
- Recordkeeping and monitoring.
- Commitment and coordination.

The importance of accountability and responsibility is stressed. Ultimate responsibility rests with the senior official of the institution. Program managers, supervisors, and employees all have key roles on which the success of the health and safety program depends.

The chapter introduces the concept that an effectively operating program depends on interaction among distinct functional parts of an institution. Five general functions are defined, and the necessary interactions among them are described. The five are

- Animal care and use.
- Research.
- Environmental health and safety.
- Occupational health.
- Administration and management.

It is suggested that the IACUC can provide helpful links among the five institutional functions.

The chapter concludes with a discussion of tasks that can aid an institution in designing an effective occupational health and safety program. The importance of hazard identification to the process is emphasized. This discussion is intended to be helpful both to institutions that have well-established health and safety programs and to institutions that are just beginning to face the task of reducing hazards to an acceptable point. For the former institutions, the discussion might help to reinforce the value of current health and safety activities and stimulate improved collaboration among programs that are supposed to protect employees. The latter institutions will find the information useful in creating a relevant health and safety program, which will require gaining an understanding of their current health and safety status, identifying existing hazards, estimating current health and safety risks and financial costs associated with them, and assessing compliance with regulations.

Chapter 3. Physical, Chemical, and Protocol-Related Hazards

Development of an occupational health and safety program depends on knowing the hazards that are present in the animal care and use setting and understanding the relative importance of those hazards with respect to the risks of occupational injury and illness. This chapter provides insight into the identification of physical, chemical, and protocol-related hazards. It describes hazards that are likely to be associated with animal care and use.

The discussion on protocol-related hazards emphasizes the responsibility of investigators to identify hazards associated with their research and to select the safeguards that are necessary to protect employees involved in the care and use of their research animals. Guidelines of the National Research Council for planning experiments with hazardous chemicals are offered as a useful approach for incorporating safety considerations into the design of protocols involving the experimental exposure of animals to toxic chemicals. Recommendations of the Centers for Disease Control and Prevention and the National Institutes of Health should be followed by investigators who are planning research activities that involve experimentally or naturally infected animals; these recommendations are briefly summarized in the text.

Chapter 4. Allergens

The prevalence of allergic reactions among animal-care workers suggests that allergenic hazards are ubiquitous in the setting of animal care and use. It is estimated that some 30% of persons with pre-existing allergic conditions, such as allergic rhinitis, might eventually develop allergy to animals. This chapter describes the types and mechanisms of allergic reactions that follow specific exposures to a variety of experimental animals. Suggestions for preventive measures and interventions are also introduced.

Chapter 5. Zoonoses

This chapter's comprehensive treatment of zoonoses will be a valuable reference for everyone interested in animal care and use. The likelihood of occupationally acquired zoonoses is much lower than it is popularly perceived to be. Knowledge of the health status of research animals and improvements in veterinary care have helped to ensure the availability of healthy research-animal populations. And exposures can be reduced even more by maintaining an awareness of zoonotic hazards and routinely carrying out appropriate hazard-control measures. The chapter presents material on zoonoses by agent category. It addresses most of the zoonotic diseases important to personnel working with laboratory animals and organizes the information according to this format:

- Reservoir and incidence.
- Mode of transmission.
- Clinical signs, susceptibility, and resistance.
- Diagnosis and prevention.

Chapter 6. Principal Elements of an Occupational Health and Safety Program

This chapter reviews the key elements of the traditional occupational health and safety program that contribute to the control of hazards and reduction of risks. Those elements constitute the scope of program activities that need to be considered in maintaining an effective occupational health and safety program. They are identified as

- Administrative procedures.
- Facility design and operation.
- Exposure control.
- Education and training.
- Occupational health.
- Equipment performance.
- Information management.
- Emergency procedures.
- Program evaluation.

The occupational-health element is treated in Chapter 7.

Chapter 7. Occupational Health-Care Services

This chapter focuses on the occupational health-care services of an occupational health and safety program. Health-care services that are appropriate for employees engaged in the care and use of research animals are reviewed. Contrary to the prevailing view in many institutions, few regulatory mandates require institutions to provide specific health-care services to employees, and such requirements that do exist are usually limited to circumstances that present substantial risk to employees. For example, OSHA's bloodborne-pathogens standard requires an institution to make hepatitis B vaccination available to all employees who handle blood, organs, or other tissues from experimental animals that are infected with hepatitis-B virus (HBV). However, the Public Health Service requirement that institutions that receive federal funds for animal research provide an occupational-health program for employees with substantial animal contact has been broadly interpreted as a mandate to provide a comprehensive array of health-care services, including physical examinations and preplacement baseline serum collection and storage. This chapter emphasizes that an adequate risk assessment must be a prerequisite in selecting appropriate health-care services for employees at risk. The following factors should be considered in performing an adequate risk assessment:

- Animal contact.
- Exposure intensity.
- Exposure frequency.
- Physical and biological hazards presented by the animal.
- Hazardous properties of the agents used in the research protocols.
- Susceptibility of the employee.
- Occupational-health history of employees doing similar work.

The health-care services that might be included in the occupational-health element are briefly described. We draw attention to considerations that are important in selecting specific services. Our discussion, however, is not meant to imply that a particular service is appropriate for all circumstances. We emphasize here, as we do throughout this book, that activities and services that are most likely to protect the occupational health and safety of employees will be judiciously based on an assessment of factors that place employees at risk for occupational injury or illness.

RECOMMENDATIONS

In writing this book, we considered several controversial issues, such as whether "substantial animal contact" was a valid indicator for determining the need for an occupational-health program. We also debated issues that were considered to lessen the effectiveness of well-intended occupational health and safety programs. The results of our deliberations are presented here in the form of specific recommendations.

Addressing Occupational Health and Safety in Animal Care and Use Programs

Many institutions that support and conduct animal research have an environmental health and safety staff that helps the institution to fulfill its responsibility to provide a safe and healthful workplace for employees. The occupational-health concerns pertaining to the care and use of research animals, however, have often not been comprehensively addressed by these institutions. In particular, the occupational-health element of an occupational health and safety program might lack focus, and its contributions to a successful program might not be well understood.

We recommend that every institution initiate a concerted effort to address the health and safety hazards and the risks of occupational illness and injury that are associated with the care and use of research animals and broaden its occupational health and safety program as necessary to reduce the risks to an acceptable level. The effort should involve the collaborative participation of people representing all institutional activities related to the care and use of research animals, including not only the animal care and use program itself, but also research, environmental health and safety, occupational health, and management and administration. Those activities should interact continually to maintain a successful occupational health and safety program. Institutions should consider the value of the institutional animal care and use committee in fostering the objectives of developing collaboration and sustaining interaction.

Institutional Commitment and Delegation of Authority

This report emphasizes the importance of interactions among many components of an institution in developing and maintaining a successful occupational health and safety program. Few criteria of success are more important than unambiguous identification of responsibility and delegated chains of authority.

We recommend that *the* senior official of an institution demonstrate personal commitment to a safe and healthful workplace, delegate clearly defined duties to those with authority to commit and direct institutional resources, and establish mechanisms for monitoring the success of the occupational health and safety program.

Risk Assessment: A Dynamic and Continuing Process

The purpose of an occupational health and safety program is to minimize risks of occupational injury and illness by controlling or eliminating hazards in the workplace. However, the need for a continuing process to review and address changes in the hazards and risks associated with new research programs, new technologies, emerging biological hazards, and the diversity of the workforce is often overlooked.

We recommend that every institution develop a multidisciplinary approach to occupational health and safety that permits the continuing evaluation of potential workplace hazards and of the risks to employees working with animals. The assessment of risk should not be limited to determination of frequency of contact, but should include the intensity of exposures, hazards associated with the animals being handled, the hazardous properties of agents used in research, the susceptibility of individual employees, the hazard-control measures available, and the occupational history of individual employees. Occupational health and safety programs should be dynamic and able to adapt to changing circumstances.

Participation in the Occupational Health and Safety Program

Many institutions limit participation in their occupational health and safety programs to full-time employees who are involved in the care and use of animals. That approach fails to acknowledge that employment status is not a relevant criterion in exposure. Students, visiting scientists, volunteers, and other nonemployees can be subjected to substantial risks associated with exposure even during brief or sporadic involvement in animal care and use.

We recommend that an occupational health and safety program provide for the appropriate level of participation of all personnel involved in the care and use of research animals on the basis of the risks encountered, regardless of their employment status.

Determining Need for Health-Care Services

Substantial contact with research animals is not a sufficient indicator of the need for health-care services. The provision of health-care services might be necessary only for particular employee groups with specifically defined occupational-health risks.

We recommend that the determination of need for health-care services be based on the nature of the hazards associated with the care and use of research animals and the intensity and frequency of employee exposure to these hazards. Overall risk assessment is key to determining this need.

Serum Collection and Physical Examinations

Serum collection and storage and physical examinations have been regarded by many institutions as typical services of an occupational health and safety program and have been applied to employees who have substantial contact with research animals. Although those services might have value for some employee groups at substantial risk for occupational illness, they are neither helpful nor cost-effective strategies for protecting the health and safety of most employees who have contact with research animals.

We do not recommend serum collection and storage as standard components of an occupational health and safety program. They have value only for employees who have substantial likelihood of occupationally acquired infection with an agent that can be monitored serologically.

We do not recommend a physical examination as the principal surveillance tool for periodic health evaluations. We recommend that a careful history based on a knowledge of workplace risks be used for this purpose. It is appropriate, however, to perform a physical examination when symptoms of work-related illness become evident during an episodic health evaluation.

2

Program Design and Management

PROGRAM GOAL

The goal of an occupational health and safety program is to prevent occupational injury and illness. The program must be consistent with federal, state, and local regulations, but the principal focus of the program should be on the control of hazards and the reduction of risks, as opposed to merely satisfying regulations. This volume is intended to raise the awareness of investigators, personnel who care for and use animals, health and safety professionals, and administrators with respect to hazards associated with the care and use of research animals and to provide some reasonable and practical approaches to minimizing health and safety risks. The principles outlined in this chapter are aimed primarily at institutions that use research animals, but they apply equally well to nonresearch animal holding, breeding, and exhibiting.

The strategies that promote health and safety in the care and use of research animals are similar to those applied generally in a research laboratory. The use of animals in research is an extension of other experimentation that occurs in the laboratory. Research animals and the procedures and techniques that attend their use, however, can present unique problems and challenges, many of which increase the hazards of experimentation. Those problems and challenges must be considered in the management of occupational health and safety programs.

DIVERSITY

A sound occupational health and safety program should recognize and reflect the wide differences in job tasks in an institution and the diversity of the personnel hired to perform those tasks. In many cases, it is difficult to identify all persons that interact directly or indirectly with animals. It is equally difficult to assign risk to each person and to determine each person's level of participation in the occupational health and safety program. Investigators, clinicians, animal-care technicians, laboratory technicians, students, workers in areas adjacent to laboratories, maintenance and custodial personnel, security personnel, and materials handlers may be included in the program. There is diversity in the health status of employees and risks associated with various work assignments performed by employees in particular job categories. Frequent turnover of employees in some job categories is inevitable and cannot be ignored in the design and implementation of a program. An accurate job description is important in identifying potential hazards.

Relevant knowledge of the people involved will vary considerably. Education and experience should be considered when assessing the need for training. The amount of and approach to training cannot be based solely on a person's educational level. For example, it is unwise to assume that someone with a graduate degree in a life science automatically requires less training in a particular aspect of the animal care and use program than someone with no college background. An experienced farm worker, however knowledgeable about a given species, might not be informed adequately on issues related to the research program and the potential hazards associated with farm-animal species in a laboratory setting. It is essential that a process be in place to assess the level of relevant training and experience of employees and to offer appropriate training at all levels.

Employees who might not be involved directly with research activities should nevertheless be included in the occupational health and safety program. For example, maintenance and custodial personnel who will have only infrequent access to animal-care areas need to be informed of the potential hazards and precautions necessary for their protection. Similarly, animal-care personnel need to be made aware of potential hazards associated with research that uses the animals under their care. Additional training might be required along those lines.

The demographics of the research community have changed. Language barriers and cultural differences must be considered and accommodated where there are people of different background and national origin, such as foreign students.

The occupational health and safety program should also recognize and reflect an understanding of the diversity of the work environment, including current facilities and potentially hazardous activities. This is a particularly important consideration during the design and construction of facilities and the renovation of existing ones. The type of research and the animal species used will influence the occupational health and safety program. For example, the use of pathogens, radioisotopes, and toxic chemicals calls for strategies different from those applied to behavioral studies in which no hazardous research agents are handled.

The use of wild-caught animals can introduce more hazards than the use of laboratory-bred animals.

BASIC CONCEPTS

Many prudent practices to protect the health and safety of workers are already widespread in institutions that care for and use animals. Such safe practices are sometimes based on common sense and sometimes on perceptions; some have been scientifically validated. An effort is made in this volume to identify practices that data have shown to be effective; when supporting data were not available, the committee suggested widely accepted practices commonly demonstrated as effective.

Programs that are intended to protect employees vary considerably from institution to institution. That is partly because institutions' needs depend on the scope of their programs. For example, containment needs for hazardous chemicals and extremely virulent microorganisms can be quite different. Institutions vary also in the sophistication of their overall health and safety programs. This volume provides guidance for all institutions in incorporating appropriate components related to animal care and use into their overall occupational health and safety program.

An effective occupational health and safety program is based on seven basic concepts:

- Knowing the hazards.
- Avoiding and controlling exposures.
- Training and education.
- Rules and guidelines.
- Consistency.
- Recordkeeping and monitoring.
- Commitment and coordination.

Knowing the Hazards

Determining the level of protection that is needed in any given situation depends on understanding the hazard in question. Defining and quantifying a hazard is sometimes referred to as risk assessment. The assessment, insofar as possible, should be based on scientific information.

In the case of infectious agents, dose-response relationships, virulence, communicability, prevalence, routes of exposure, shedding patterns, stability, and availability of prophylaxis and therapy are important considerations. For chemical agents, one has to know about toxic doses, stability, form (liquid, gas, or solid), type of toxicity (irritation, corrosion, carcinogenicity, narcosis, lethality, etc.), severity of reaction, mode of action, and metabolic products. The main sources of information for risk assessment are the scientific literature and professionals and consultants with unpublished field experience. Those sources might fail to provide the necessary information, so additional research or increased caution is sometimes warranted.

Experience has shown clearly that undetected hazards pose a major problem. Many undetected hazards are not related to the intended use of animals in a laboratory. For example, an animal might carry a human pathogen into a laboratory. Unfortunate incidents have occurred when animals harboring zoonotic diseases, such as Q fever and lymphocytic choriomeningitis, have been used in research laboratories (Bowen and others 1975; CDC 1979; Hotchin and others 1974; Jahrling and Peters 1992; Spinelli and others 1981).

Avoiding and Controlling Exposures

It is common sense that it is better to avoid a hazard than to deal with the consequences of exposure to it. Measures related to this principle include training, work practices, containment equipment, personal protective equipment, control of access to hazardous areas, and use of purpose-bred animals. Safety measures should be implemented in advance rather than after a problem emerges. Although reducing risk to employees is the primary goal of an occupational health and safety program, it should be recognized that it is impossible to eliminate risk.

Training and Education

Once a hazard is known, this knowledge must be communicated to animal care and use employees most directly involved and other employees (such as janitorial and maintenance workers) who might be at risk of exposure. Employee training begins with orientation immediately after hiring. Standard operating procedures should include methods for performing duties safely. New employees should be carefully instructed in those procedures by an experienced co-worker before assuming duties independently. Laboratory procedures can be reinforced with signs and posters. Periodic meetings to encourage safe work practices are advisable, and safety newsletters and electronic bulletin boards sometimes can be beneficial in keeping employees updated on changes. An institution has a crucial role in ensuring that its employees remain both well informed of relevant health and safety information and proficient in the use of safe practices.

Rules and Guidelines

Rules are necessary to ensure safety in the workplace. Rules governing the training of personnel, adherence to work procedures, use of disinfectants and decontaminants, access, waste disposal, use and maintenance of equipment and safety devices, emergency procedures, reporting of accidents and exposures, and personnel behavior (smoking, eating, and hand-washing) should be rigidly enforced. Rules are "musts," whereas guidelines are recommendations and suggestions that allow for some judgment. For example, "No smoking, eating, or drinking in the animal room" would be a rule, whereas "The recommended use of chemical restraint before the use of hands-on procedures involving aggressive animals" would be a guideline.

Consistency

Consistency is essential to the success of an occupational health and safety program, including consistency in rules, enforcement, and application to all workers. Lack of consistency can undermine a program. For example, if employees are expected to wear masks and gowns to enter specified animal rooms, both supervisors and animal care and use staff should wear masks and gowns. When higher-level personnel ignore rules, it sets a bad example. However, too-rigid safety rules, at times considered unreasonable by employees, can undermine the credibility of a program.

Recordkeeping and Monitoring

Developing and maintaining records is essential in an occupational health and safety program. It might start with a medical history of each employee to discover any facts that would bear on the general susceptibility of the employee to injury or illness. Reports of accidents, exposures, and work-related illnesses are absolutely necessary and sometimes required by law. Other forms of recordkeeping can provide useful information for monitoring safety programs and identifying deficiencies.

Commitment and Coordination

Commitment to safety must be a feature of an organization from top to bottom. Even the best safety program will fail if employees ignore the rules. The hierarchy of management must be committed if a safe attitude is to be instilled in workers.

Animal facilities are rarely autonomous organizations; coordination is required among administrators, research scientists, veterinarians, technicians, and maintenance workers. Every person's role should be clearly defined because safety programs can fail if responsibilities are diffuse and not well understood.

ACCOUNTABILITY AND RESPONSIBILITY

Different parts of the overall occupational health and safety program are necessarily managed by different people. The level of responsibility and accountability for the design of the occupational health and safety program and for the program components should be well defined and can be divided into four levels: the institution, the program managers, the program implementors, and the individual employees.

The Institution

The institution, represented by its senior official (or an authorized body), has ultimate responsibility for providing a healthful and safe work environment and must have a commitment to that goal. The senior official must

- Understand the issues.
- Provide guidance.
- Establish and support institutional policies.
- Have authority to provide necessary resources.
- Bring together program managers and implementors.

Demonstrated actions of the senior official are essential to the success of the program. The senior official also makes assertions to regulatory agencies regarding compliance and must be confident that these assertions are valid and backed by appropriate documentation.

The senior official is accountable for health and safety in the work place. This official must reasonably delegate or assign responsibility and authority for the program components to other appropriate persons. The official must have an adequate understanding of both technical and management issues and should be routinely advised in matters related to the program. Either a person, a task force, or a committee might be effective in addressing the complexities of designing and implementing a sound occupational health and safety program, but throughout the process the lines of authority must be clear, with all participants understanding the communication and interaction needed to attain the objectives. A performance-based system or process, which focuses on the desired outcome of worker health and safety, can help an institution to address the many issues involved.

Program Managers

An effective occupational health and safety program depends on the involvement and commitment of program managers at all levels. Key managers will be those who have specific expertise in health and safety issues or who will be charged with and have the authority to implement and enforce components of the program. The program managers should include, if appropriate, the following:

- Health professionals.
- Safety professionals.
- Veterinarians.
- Animal-facility managers or supervisors.
- Research directors and scientists.
- Laboratory supervisors.
- Human-resource and finance personnel.
- Legal advisers.
- Environmental experts.
- Facility engineers.

It is imperative that research scientists, who direct experimentation, participate in the design, implementation, and management of the occupational health and safety program.

Program Implementors

Responsibility for implementation resides at the supervisory level. Training is a key function of an implementor's responsibility. Training should emphasize the active and preventive nature of effective safety programs. Training programs should provide adequate information about the risks involved and preventive measures available. Implementation also involves providing appropriate equipment for personal protection, providing appropriate facilities, and ensuring compliance of subordinate staff with established procedures and practices.

Employees

For the purposes of a sound occupational health and safety program, an employee may be defined as any person who might be at risk from any activity within the institution that involves or is associated with the care and use of animals in research. All employees must share responsibility for their own health and safety and for the safety of those around them. All must work so as to protect themselves and others and incorporate safety into day-to-day activities. That requires that employees comply with rules, follow established standard procedures, report injuries, and be generally active in demonstrating safe work practices. Employees at all levels should be involved in program design and improvement.

Apart from regular employees, consideration should be given to the health and safety of students, part-time and temporary employees, full and part-time contractors, visitors, and volunteers.

INSTITUTIONAL ACTIVITIES AND THEIR INTERACTIONS

An institution that uses animals in research is responsible for five main activities:

- Animal care and use.
- Research.
- Environmental health and safety.
- Occupational health.
- Administration and management.

Interactions among these activities are important for implementation and maintenance of an effective occupational health and safety program. The central focus of the health and safety issues discussed in this document is the care and use of animals in research, which includes the established animal care and use program and the institutional procedures for review and monitoring of animal use. It involves mainly a program manager, who is usually a veterinarian; the animal-care staff; and the institutional animal care and use committee.

Research involving animal use is conducted by investigators and technicians in the research laboratories and in the animal facility. Scientists' research objectives are directly supported by the animal care and use program.

The environmental health and safety activity provides technical services that assist the institution in carrying out its responsibilities associated with health and safety; it involves people who have expertise in chemical safety, biological safety, physical safety, industrial hygiene, health physics and radiation safety, engineering, environmental health, fire safety, and toxicology. Included in this activity are programs to collect, transport, and dispose of hazardous waste; manage responses to emergencies; monitor regulatory compliance; and provide training support and technical assistance. Occupational health involves primarily health-care professionals, including physicians, occupational health nurses, and specialists required to assess potential health risks and manage the care of employees who have acquired an occupational injury or illness; it is often organizationally connected with the environmental health and safety activity.

The administrative and management activities include involvement of the senior official and program managers and other human-resources, finance, risk-management, and property-management personnel.

This section provides examples of these interactions among the five general activities. The discussion is intended to help an institution to identify potential interactions that will make it possible to carry out an effective occupational health and safety program (Figure 2-1).

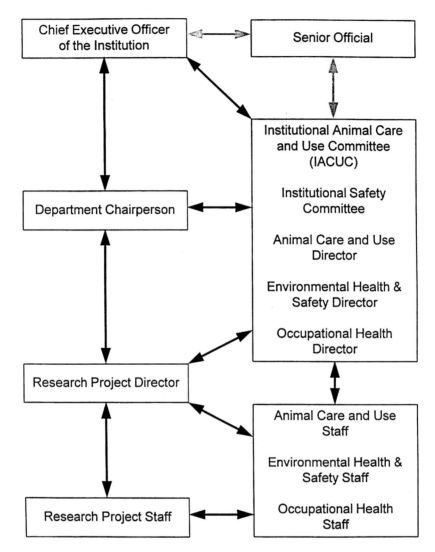

FIGURE 2-1 Pattern of interactions.

Animal Care and Use

For purposes of this document, an occupational health and safety program is built around animals and their use. Most institutions have an institutional animal care and use committee (IACUC). By virtue of the IACUC's responsibility for review and monitoring of animal use, it can help to provide links among many of the institutional functions related to health and safety.

Review of Proposed Animal Use

A critical component of any program is the identification of potential hazards. Review of proposed animal use by the IACUC can appropriately include requests for information on the potential hazards involved in the proposed research protocols. That does not imply that the IACUC must conduct the review and assessment of potential hazards. But the committee is charged with ensuring or being assured that appropriate review is taking place. Hazard review can be accomplished by obtaining the advice or approval of other activities, committees, or responsible persons. For example, approval by the radiation-safety committee might be required before a study involving radionuclides is approved by the IACUC. Identification of potential hazards associated with research can also result in identification of concerns related to the husbandry of animals. Impacts of research on activities related to animal care, such as bedding disposal, are often overlooked. Additional training or hazard communication might be necessary for animal-care personnel in subjects not routinely encountered.

The staff veterinarian should be closely involved in the planning and review of proposed animal uses. The IACUC and the veterinarian should specifically consider the potential for zoonotic disease and other potential hazards associated with the species involved. Consideration might involve the requirement for specific training, preventive measures, or special health monitoring. Communication between the veterinarian and the investigator during the planning phase can address all those concerns before a proposal is presented to the IACUC for review and approval. In addition, changing circumstances associated with the introduction of new research projects can be considered.

The protocol-review process should provide an opportunity for identifying the personnel who will be involved in a research activity and therefore the persons potentially exposed to the associated hazards. It can stimulate close communication with the investigator regarding the necessary training and experience of research staff. Involvement of the human-resources function during the hiring process can be helpful in identifying animal users.

Monitoring of Animal Use

Monitoring of animal use is a continuous process. Hazard awareness and attention to safety should be the responsibility of every employee; this responsibility includes following established rules and procedures, reporting injuries, identifying hazards, and demonstrating safe work practices at all times.

Laboratory managers and supervisors must monitor compliance with established procedures and periodically review their appropriateness. Training of personnel is an important part of this process.

The IACUC is required to conduct periodic reviews of the animal care and use program and to inspect the animal facility and animal-use areas. It is appropriate during review and inspection to consider all the components of the occupational health and safety program, including the research in progress. The facility inspection should include review of compliance with safe work practices and standard procedures, as well as the adequacy of the facility and equipment.

The veterinarian and animal-care staff are in a unique position to monitor animal use continuously. Daily contact with the animals gives them a direct perspective of the use of the animals and, often, of the effectiveness of and compliance with health and safety procedures.

Research

The importance of involvement of research investigators and their staff in the design, management, and implementation of an effective health and safety program has been mentioned. Interaction with the veterinarian and the IACUC during project planning has become an expected and useful kind of communication. Compliance with radiation standards and communication with the radiation-safety officer have long been practiced. Approvals for chemicals, infectious agents, or recombinant-DNA use might also be required. Many aspects of the health and safety program rely on active participation of workers in the research laboratories, not only in the animal facility.

Environmental Health and Safety

Responsibility for coordinating the occupational health and safety program often is delegated to the environmental health and safety staff. This staff should be responsible for, or involved in, the establishment of prudent practices that comply with Occupational Safety and Health Administration (OSHA) standards. The staff should participate with occupational-health professionals in the assessment of risk. Training and hazard communication should involve all appropriate employees. Information on zoonoses and specific animal-facility hazards should be provided, and this requires coordinated efforts among many activities listed in this section.

The radiation-safety committee and the radiation-safety officer are responsible for establishing programs that ensure the proper use of ionizing and nonionizing radiation. The use of radionuclides in an animal-research facility—subject to environmental health and safety review—involves a coordinated effort between the radiation-safety committee, the research staff using the materials, and the animal-care staff. In most cases, the animal-care staff has little training in or understanding of radionuclide use and the associated hazards. Special training for animal-care staff, at a level of understanding necessary to provide safe routine husbandry, is required; it could be coordinated by the facility veterinarian (to identify persons involved in the care of the animals) and investigators (to determine the nature of the hazard and the potential risk to animal-care personnel).

Waste management should also be under the purview of the environmental health and safety staff. In association with the return of animals to the animal facility and their maintenance after experimental exposure, procedures for husbandry, waste disposal, and monitoring should be well defined, communicated to all personnel, and followed.

Monitoring the effectiveness of the occupational health and safety program depends on analysis of injury and accident data, which are generally managed by the environmental health and safety staff and the occupational health staff. Obviously, compliance of individual employees with injury-reporting requirements is essential to this effort. Close interactions among these health and safety professionals will make these retrospective studies more meaningful.

Occupational Health

Identification of new employees at risk and rapid evaluation for inclusion in the occupational health component of the occupational health and safety program is essential to ensure their protection. Close involvement of the occupational health staff with the human-resources and legal staffs of the institution is desirable. Preplacement health evaluations and discussions with health professionals can provide an opportunity to establish a potential exposure profile and train and inform new employees of institutional policies and requirements.

Occupational health physicians and nurses can be most effective if closely involved in or provided with a good understanding of the specific nature of potential and actual occupational exposures. The occupational health staff should be involved in the assessment of workplace hazards and risks; this activity is generally the responsibility of environmental health and safety staff but necessitates the participation of health professionals, the veterinary staff, and the scientists conducting the research. Close links to the animal care and research activities are essential to an effective occupational health activity.

Administration and Management

Ultimately, the quality of an institutional occupational health and safety program depends on the support of management. Involvement of financial-management personnel for budget planning and resource allocation is needed particularly when a comprehensive program is being implemented or equipment and facilities appropriate for the conduct of research are being provided. In addition, the periodic review of existing and desirable program elements by persons with control of institutional resources is helpful. The human-resources function can constitute an effective communication link to the research program through involvement in job classification, preparation of job descriptions, and information exchange during the hiring and orientation process.

MANAGEMENT STYLE AND STRUCTURE

A most-important element in managing a successful occupational health and safety program is a favorable attitude toward promoting safe working conditions. There are several ways to generate and encourage such an attitude. The foremost is to involve representatives of all activities in the task of developing the occupational health and safety program. The concept of safety must be presented in a cooperative spirit. The idea is to assist in reaching the institution's goals without sacrificing safety. Safe methods should be developed in a helpful manner and not with the threat that something will be done in some specific way or not at all.

Within the organizational hierarchy, another management style that fosters cooperation is leading by example. The converse of that is particularly damaging to a safety program; that is, if the supervisors ignore basic safety rules, they cannot expect their employees to behave differently.

Rewarding employee compliance and admonishing those who break the rules is essential in communicating the importance of safety. Safety awards provide individual recognition and reinforce employee dedication. An employee who is apathetic or indifferent about safety should be counseled that such behavior jeopardizes everyone and will not be tolerated.

There is no prescription for structuring an institutional occupational health and safety program. The size and complexity of a given institution influence the overall structure of its program. Some institutions centralize their occupational health and safety program; in others, the programs are more diffuse. In either case, it is important that all the elements of the program be covered and that responsibilities be clearly assigned.

GETTING STARTED

Each institution has its own organizational history and culture. The task of designing an occupational health and safety program for employees involved in the care and use of research animals will benefit from having an "institutional champion" to orient and guide the task group through the institutional maze; it is essential for defining the organizational boundaries of the five activities described earlier and learning helpful strategies for establishing necessary interactions among them.

The availability and effectiveness of all elements of a program will depend on one absolute consideration: the senior official of the institution must be genuinely and openly committed to maintaining an occupational health and safety program. A sincere commitment will ensure the availability of the support and resources necessary to enable the program to be operated effectively. Trying to develop an occupational health and safety program without the support of institutional leadership is a losing proposition. Only the institutional leadership can identify the appropriate chains or networks of communication and authority within the institution.

Priority List

Many programs stumble by trying to initiate too many program components at once. It is useful to establish a priority list based on existing hazards and the current occupational health and safety status of the institution.

Institutions typically choose to address the exposures that are causing the greatest current costs. Costs can be measured objectively, such as payments for worker compensation, or measured subjectively, such as unwanted mass-media attention, regulatory citations, or worker alarm over perceived hazards (independent of measurable risk). In the absence of an occupational health and safety program, relatively small and otherwise manageable problems can cause great institutional disruption.

Once acute problems have been addressed, a priority list can be compiled on the basis of assessment of the magnitude and severity of identified risks. An institution might choose first to address the most-common or most-severe risks, or a combination of both, and then to identify exposures that, over time, would result in an unacceptable cumulative risk. Preparing for accreditation visits or OSHA inspections can focus institutional attention on issues of particular interest to such agencies.

There is a hierarchy of control and prevention strategies. Primary prevention of occupationally acquired injury or illness is achieved by controlling or eliminating hazards, and the quality and effectiveness of an institution's occupational health and safety program will depend on how well resources are distributed to provide for and promote hazard-control strategies. Secondary prevention (premorbid case detection) and tertiary prevention (case-finding and disease management) are less desirable as means of controlling occupational health and safety risks.

Hazard Identification

Identification of hazards is a challenge in all workplaces; animal care and research facilities are no exception. No databank, book, or journal will definitively identify all the hazards in the workplace. Identification of hazards is a responsibility of everyone: supervisors, managers, investigators, and other employees. Many hazards are readily apparent. For example, lifting heavy animal cages and then twisting to put them onto a conveyor belt as it enters the cage-washing unit obviously constitutes a hazard. Others might require special knowledge to identify, such as ultraviolet light sources in an area where chlorinated solvents are used (the resulting phosgene production presents a hazard). Some hazards, like allergens, are ubiquitous but complicated by individual susceptibilities and nonoccupational exposures. The challenges are to identify as many hazards as possible and to keep an open mind to new ones.

An efficient way to identify hazards is to have an environmental health and safety professional who is trained in the recognition, measurement, and control of workplace hazards perform a walk-through review of the animal facility and the laboratories of investigators who conduct animal research. The quality of the review would be enhanced by the participation of a knowledgeable person from both the animal-care program and the research program. The review should be conducted when animal care and use are in progress. The reviewer should be attentive to the worker, the environment, the protocols, and the equipment. Operations that are being performed and how the employees are performing them should be carefully observed. Obvious signs of exposure—such as the presence of dust, nasal or eye irritation, chemical odors, and the accumulation of chemicals on work surfaces—should be noted. Observations should be compared to the expected performance of a well-trained worker operating properly designed equipment in a clean and safe work environment. The reviewer should understand the operations well enough to construct scenarios that might result in personal injury. The purpose of the review should be discussed openly with the workers, and their observations and viewpoints should be sought. Workers are an indispensable source of information concerning the hazards that are associated with their work and should be encouraged to report hazards observed in the workplace either to their supervisor or to the environmental health and safety office.

Many institutional data sources contain hazard information, e.g., accident reports, reviews of experimental protocols submitted by investigators, manufacturers' safety bulletins, safety reports prepared by labor unions, safety-committee reports, job-safety analyses, safety-audit reports, health and safety consultants' reports, Material-Safety Data Sheets (MSDSs), and chemical inventories. Not all those sources are available in all workplaces. The sources that do exist should be consulted because they can provide unique site-specific information.

The experience of other institutions can be helpful in identifying hazards. The Institute of Laboratory Animal Resources is an invaluable repository of relevant information and institutional contacts. Professional associations and societies are among the best sources of information on animal-related health and safety issues. Those and other professional organizations hold conferences at which health and safety issues are discussed. Much of the value of such meetings results from the focused nature of the presentations and their application to specific operations. Professional societies are excellent sites for developing networks of experts in health and safety programs.

Government agencies also have knowledgeable personnel who can be of great assistance. The relevant agencies include the National Institute for Occupational Safety and Health (NIOSH), the Occupational Safety and Health Administration (OSHA), the Centers for Disease Control and Prevention (CDC), the

National Institutes of Health (NIH), the National Animal Disease Center (NADC), the Agricultural Research Service (ARS), and the Animal and Plant Health Inspection Service (APHIS).

Given that a hazard is present in the workplace, the next question is, "How important is this hazard?" In answering this question, consideration needs to be given to the number of people who are exposed to the hazard, the potential effect of the hazard on the people, and the magnitude of the exposure. Exposure to a carcinogen from a skin-painting experiment would be considered a more-serious hazard than exposure to ammonia gas emanating from bedding materials. For the ranking process to be effective, however, it must take into account the frequency or probability of some consequence. For example, an exposed electric circuit on a cage-washing machine presents a hazard of electric shock. If, however, the unit is in a locked room and not used, the risk of an electric accident is very low. If the unit is the primary cage-washing machine and is used daily, the chance of an electric shock is much greater. In that simple example, the consequence (the electric shock) and the hazard (exposed electric contacts) are the same in both scenarios, but the likelihood of the consequence is different. Risk is a measure of the likelihood of a consequence, whereas hazard is the inherent danger in a material or system. Ranking of hazards on the basis of the characteristics of the consequence and the likelihood of the consequence enables an institution to understand the occupational health and safety risks in its animal-care and research programs and to plan appropriate risk-reduction strategies. The principal objective of an occupational health and safety program is to reduce to an acceptable level the risk associated with using materials or systems that might have inherent danger. That is accomplished by controlling or eliminating hazards.

The actual injury and illness experience within an institution among employees who are involved in the care and use of research animals is a key indicator of the presence of workplace hazards and can be used to estimate occupational risk and help to rank the importance of existing hazards. For example, animal bites and kicks are common but rarely life-threatening; back injuries are common and tend to be of greater severity and to lead to greater overall costs; laboratory-animal allergy is common and has a wide range of severity, from mild rhinitis through chronic asthma to life-threatening asthma or anaphylaxis; dermatitis is less common and rarely severe but can reduce barriers to other hazards like infectious agents; and B-virus infection is uncommon but life-threatening.

Information about current occupational health and safety status in a particular institution can be gathered from several key sources in that institution, as shown in Table 2-1.

- Worker compensation. Worker compensation is the insurance system maintained by an institution to cover the medical costs and replace lost wages of workers with work-related illness or injury. The worker-compensation carrier typically can provide summaries of costs and lost days by medical diagnosis.

TABLE 2-1 Sources of Information About Worker Health and Safety

Information source	Who has information	Relevant institutional data	Reference data
Worker-compensation			
• First Report of Injury or Illness	Administration and management	Number of reports	Institutional trends
• Insurance companies	Institution's insurance carrier	Claims experience: direct medical care costs and lost-time costs	Industry and company special reports, institutional trends, compensation agency reports, Federal Bureau of Labor Statistics Supplementary Data System (SDS)
OSHA 200 log "OSHA recordables"	Environmental health and safety office, administration, and management (human resources)	Number of entries	Institutional trends, OSHA "compliance database," National Safety Council "Accident Book," Bureau of Labor Statistics annual summary of injury and illness
First-aid log	Institution	Number of entries	Institutional trends
Occupational health log	Occupational health office	Periodic visits (participation rates), episodic visits	Institutional trends
Adverse-reaction reports	Environmental health and safety office	Number of reports	Institutional trends

Costs and lost days per case can also be summarized to provide an index of the severity of an illness or injury. This information is extremely helpful in identifying hazards that cause the greatest adverse health effects in workers. Most institutions choose to address those hazards quickly by putting into place adequate control strategies to avoid future compensation costs. The Federal Bureau of Labor Statistics receives data from 35 states that categorize and report injuries and illnesses that qualify for worker compensation. The resulting database is called the Supplementary Data System (SDS) and is available for public use. It can be searched to identify injuries and illnesses that have occurred at similar institutions and to discover the injury and illness experience of a specific group of workers. For example, the most recent publicly available data were obtained for the year 1986 on all injuries and illnesses that qualified for worker-compensation payments in California for the standard industrial classification code (SIC) 0740 (veterinary services). The data were analyzed to determine which kinds of workers were most likely to be injured and which types of injuries were most common. A total of 74 compensable injuries and illnesses were reported in this SIC. Of the 34 injuries that occurred among workers classified as "animal caretakers," 54% were due to cuts, punctures, and bites by animals and 30% to overexertion due to lifting (SDS 1994). Information of that type is helpful in developing an understanding of workplace hazards and in grouping workers on the basis of risk. Because worker-compensation insurance carriers are paid by employers, they can also provide historical data on an institution's operations and analyses of trends to help reduce injuries.

- First Report of Injury or Illness. These reports are used to inform the worker-compensation insurance carriers of the occurrence of a probable occupational injury or illness. A report generally describes the injury or illness and identifies causative factors that might be immediately apparent. Reports can be acceptable alternative records to the Supplementary Record of Occupational Injury or Illness (OSHA Form 101) that employers are required to maintain (BLS 1986).

- OSHA 200 log. Most institutions are required by law to maintain a log of work-related illness and injury, commonly known as an OSHA 200 log (US Dept. of Labor). Injuries are defined as incidents that are instantaneous, such as a bite, kick, or needlestick. Illnesses are defined as conditions arising from noninstantaneous events, such as carpal-tunnel syndrome, animal allergies, and dermatitis. An injury is recordable if it results in the death of an employee, loss of consciousness, lost work time, placement on restricted duty, or treatment other than first aid. All recognized occupational illnesses are recordable. Lost time need not be incurred for an event to be recordable. The log does not provide a comprehensive summary of all work-related events in that minor injuries might not meet any criteria for being recordable. The same single entry is required for a minor or severe illness and for a recordable injury; hence, the log does not show the intensity or cost of an individual illness or diagnosis. However, a Supplementary Record of Occupational Injury or Illness (OSHA Form 101) must be prepared and kept by employers for each OSHA 200 log entry; this record contains detailed information concerning the injury or illness in question and can be helpful in investigating their sources.

- First-aid log. First-aid logs are generally maintained by supervisors at the worksite. They are a useful source of information about minor occupational illnesses and injuries that are treated outside the occupational health unit; for example, workers with eye injuries are usually sent directly to an ophthalmologist or emergency service, and the health unit might be unaware of their occurrence. They are also a source of information about nonrecordable minor injuries.
- Occupational health log. Institutions that have health units or provide other health services to employees will have available other information useful in assessing occupational risks of employees. Periodic and episodic visits to these clinics are the source of this information. Periodic visits are routine, scheduled visits for preventive care (e.g., immunizations or surveillance evaluations). The number of periodic visits at an employee health unit is determined by the number of workers who are eligible to participate in a medical-surveillance program, the magnitude of risks, and the rate of participation in established surveillance programs. Surveillance visits can yield information on the prevalence of a condition in the worker population (e.g., the number of workers with animal-related allergy) and on the frequency of specific risks (e.g., a practitioner can ask a worker to estimate the number of times that he or she has been bitten in the preceding year). Episodic visits are nonroutine, unscheduled visits that are needed because of the occurrence of known or suspected work-related illness or injury. Episodic visits reflect the occurrence of cases of sufficient severity to require health services.
- Adverse-reaction reports. Adverse-reaction reports identify symptoms or occurrences at the worksite that suggest increased risk from a hazard. They are typically kept by the environmental health and safety office but reviewed by the occupational health unit to determine whether a medical evaluation is needed. Each worksite determines the threshold for reporting an adverse reaction. Accident reports are included in this category.

Another source of information on hazards in the workplace is the record of citations issued for violations of OSHA regulations. It is useful to know what worksite conditions are commonly monitored by OSHA and what has been the basis for citations at other institutions. Institutions with established occupational health and safety programs are a good source of clarification. Information on hazards cited by OSHA is collected by OSHA and maintained in the Compliance Database, which contains information on all the hazards and company data observed by OSHA during inspections for which a citation was issued. Although it is limited to compliance data, it is useful because it provides a nationwide view of the safety-compliance issues in similar operations. Data are available on the organization, the type of citation, and the number of workers exposed. To obtain information in the Compliance Database, request it from one of the OSHA regional offices.

Work Plan

As an occupational health and safety program is developed, periodic meetings of representatives of animal care and use, research, environmental health and safety, occupational health, and administration and management who will become involved in proper implementation are important. The meetings will promote the necessary coordination of activities. It is easier to establish or expand an occupational health and safety program if diverse program elements are represented in one room as the broad outlines of the work plan are developed. The frequency of the meetings will depend on the magnitude and complexity of the task.

Measures of program effectiveness should be established and agreed on by the group. Measures of program effectiveness could include reductions in chemical-exposure levels, specific injuries or illnesses, damaged-material costs, loss of work because of damaged equipment, and program costs per covered employee. If the program has clear goals, measuring its effectiveness can be straightforward.

Plan for Resource Development

Diverse resources are needed for the successful implementation of an occupational health and safety program. Administration and management often need to be sold on the value of the program; this is sometimes a challenge because it is difficult to measure avoided costs.

A workforce that is aware of workplace hazards and proficient in practices for their control is perhaps the greatest resource that an institution can develop. It is appropriate to seek resources to train workers in the recognition and avoidance of hazards and the conduct of safe work practices. Resources should also be sought to correct workplace conditions that require workers to engage in repeated or cumbersome protection practices for safety.

Hardly any problem encountered in laboratory-animal research is new. Time and money can be saved by seeking the advice of those who have dealt with similar problems and found suitable solutions. Formal or informal consultation can provide experience and perspective on issues for which programs must be developed in a setting of uncertainty. One of the primary functions of a consultant is to inform an institution of occupational health and safety facts that provide clear direction for program development and advise of gaps in occupational health and safety facts that must be filled by considered judgment.

The work of an occupational health and safety program is accomplished by people with a detailed knowledge of the particular workplace, its research activi-

PROGRAM DESIGN AND MANAGEMENT

ties, and the institutional history of success and failure with control strategies. It is difficult for consultants to acquire and maintain this knowledge base.

An occupational health and safety program is such a multifaceted enterprise that it is unlikely that a few persons will have all the expertise required. Specialists might have to be added to the environmental health and safety staff. In institutions that lack an occupational health staff, necessary service can be obtained by contract. Such services can be in a medical center or a free-standing clinic or can be contracted to be established on the premises. It is essential that administrative oversight and responsibility be assigned to someone within the institution. Continuity of records and continuity of services are especially important if contract services are used.

Plan for Evaluation and Update

As a program develops, there can be an apparent increase in the frequency of occupational health and safety problems as case-finding increases. That is often unsettling for the administration and can mistakenly be taken as a sign of program failure. More commonly, it is a sign of a successful program, and costs of the occupational health and safety program decrease as the severity of conditions found decreases because of prevention and early diagnosis.

3

Physical, Chemical, and Protocol-Related Hazards

The diversity of physical, chemical, and protocol-related hazards associated with animal research is tremendous. Animals bite, scratch, and kick; moving bulky animal cages can result in sprains and strains; and electricity, machinery, and noise can cause injury. Chemicals are ubiquitous in the laboratory and animal room environments; chemicals are used to disinfect and clean surfaces, anesthetize animals, and process tissue samples. Research protocols can introduce toxic chemicals, human pathogens, or radioactive materials into animals, and these agents can enter the waste stream of the animal facility. This chapter provides a brief review of specific physical, chemical, and protocol-related hazards that are commonly observed in animal care and use programs.

PHYSICAL HAZARDS

Animal care and use by their very nature present many situations that require safe practices to protect workers from physical hazards. The hazards of bites, kicks, and scratches are associated inevitably with most laboratory animal contact. A survey of animal-related injuries among veterinarians indicated that 35% required sutures for lacerations during their career. Working with heavy animals and equipment, such as metal cages, can stress muscles and joints. The potential for wet floors in animal rooms and cage washing areas increases risks of slipping and falling. Workers can also be exposed to physical hazards that are commonly found in the research environment, such as flammable solvents, ultraviolet radiation, ionizing radiation, pressure vessels, noise, and electric shock. The physical hazards selected for discussion in this section present the highest potential for causing serious harm and are likely to be present in most animal facilities.

Animal Bites, Scratches, Kicks, and Related Hazards

Bites, scratches, and kicks are ubiquitous hazards associated with laboratory animal contact. They are largely preventable through proper training in animal-handling techniques. People working with large domestic animals might sustain crushing injuries when the animals kick, fall, or simply shift their body weight.

Personnel should be aware of environmental factors, as well as factors intrinsic to the animal, that can precipitate a traumatic event in a research animal facility. Several factors need to be considered in work with animals (Grandin 1987). Animals respond to sounds and smells as people do; they also hear, smell, and react to things that people might not detect. If an animal hears a high-pitched sound, it might become frightened. Such situations can result in an unexpected response that results in injury to the animal handler. Many animals have a "flight zone": approaches by another animal or a person cause an attempt to escape. Being aware of an animal's flight zone will help to avoid injuries. Many animals, including monkeys and livestock, are social and show visible signs of distress if isolated from others of their kind. Knowledge of animal behavior is important in reducing risks.

Inappropriate handling can induce discomfort, pain, and distress, provoking an animal to inflict injury on its handler. Personnel should review educational materials pertinent to safe animal-handling techniques (Fowler 1986; Kesel 1990) and should have supervised instruction before undertaking new animal-handling procedures. The institution should be prepared to evaluate the causes of any injuries that result from newly adopted procedures. The injured persons should participate in this evaluation.

Special attention should be given to the training of personnel involved in the handling and restraint of nonhuman primates. In addition to posing a bite and scratch hazard, nonhuman primates can be challenging and difficult to handle safely because of their great strength, dexterity, intelligence, and tenacity. Unsuspecting personnel have been injured when nonhuman primates have grabbed and pulled neckties, loose-fitting laboratory coats, or long hair, and some individual great apes have been known to throw their feces. When it is compatible with the experimental conditions of animal use and the clinical condition of the particular animal, consideration should be given to chemical immobilization of nonhuman primates to facilitate the ease of handling them and to reduce the risk of injury of personnel. Personnel who work with nonhuman primates should wear face shields and other protective garments and equipment appropriate for the circumstances and species involved.

In a survey of animal bites among the general population, dogs were the species most commonly involved, with cats and rodents second and third (Moore and others 1977). Comparable data on bites in animal facilities are not available, but rodent bites probably predominate because of the large number of rodents used and the broader exposure of personnel to them.

Animal bites, especially those by rodents that inflict little tissue damage, are sometimes considered inconsequential by personnel who are unfamiliar with the host of diseases that can be spread by this mechanism and the complications that can result from wound contamination by the normal oral flora of the animals involved. Personnel should be alerted to the need to ascertain their current tetanus-immunization status, seek prompt medical review of wounds, and initiate veterinary evaluation of the animal involved if it is warranted. Rabies, B-virus infection, hantavirus infection, cat-scratch fever, tularemia, rat-bite fever, brucellosis, and orf are among the specific diseases that can be transmitted by animal bites with profound consequences (covered in more detail in Chapter 5).

The early initiation of antimicrobial therapy for all animal bites that are not trivial appears warranted because there is a high probability of wound contamination with a potential pathogen. That approach will limit the progression of a localized infection and avert the more serious complications of wound infection, which could include cellulitis, abscess, septic arthritis, tenosynovitis, osteomyelitis, sepsis, endocarditis, and meningitis. If infections do not respond to therapy, additional microbiological studies that encompass unusual and fastidious organisms should be pursued. Fungal agents should not be overlooked as possible wound contaminants; the transmission of blastomycosis to humans by dog bite has been reported (Gnann and others 1983).

A wide variety of poisonous and venomous reptiles (Russell 1983), marine animals (Halstead 1978), and arthropods (Biery 1977) might be maintained in the laboratory or animal facility for research or instructional purposes. Institutions that host these uncommon research animals have a special obligation to perform a comprehensive review of safety precautions to ensure the security of animal housing and the appropriate training of personnel who are involved in their care and use. Institutions also should have a plan for the immediate delivery of definitive medical care in response to envenomation, including the use of antivenin if available. Many types of envenomation cause massive tissue destruction that predisposes a wound to secondary bacterial infection and indicates a need for treatment with tetanus toxoid and antimicrobial therapy (Goldstein 1990a; Sanford 1985).

Sharps

Sharps are ubiquitous in animal care. Needles, broken glass, syringes, pipettes, scalpels—all are commonly used in animal facilities and laboratories. Controls include installing puncture-resistant and leakproof containers for sharps at critical locations in the facilities. Workers should be trained to handle and dispose of sharps safely. Improper disposal of sharps with regular trash can expose custodial staff to puncture wounds and cuts and potentially to exposure to infectious agents and hazardous chemicals. Many states and some municipalities have regulations that specify how to dispose of sharps; these regulations should be checked to ensure that disposal practices are in compliance.

Special care is required in the use of needles and syringes to avoid needlestick injuries. This hazard presents a substantial risk for occupationally acquired infection in inoculating or drawing blood from laboratory animals (Miller and others 1987). Appropriate restraint or sedation of animals during procedures entailing the use of sharps decreases the risk of sharps injury to workers.

Flammable Materials

The National Fire Protection Association (NFPA) has classified fires into four types according to the character of the flammable or combustible materials. Class A, B, and C fires involve general combustible materials (such as wood, paper, and cloth), flammable gases and liquids (such as oil and paint), and electric equipment, respectively. Class D fires involve such combustible metals as magnesium, sodium, and potassium. Class A, B, and C materials are found in all animal care facilities. Common combustible materials in Class A fires found in animal care facilities include animal bedding, paper gowns, plastic animal cages, paper towels, and laboratory wipes. Class B flammable solvents might be used in painting animal care rooms, cleaning floors and surfaces, sterilizing equipment, administering anesthesia, and performing laboratory analyses of tissues. Common Class C materials include lighting, wet vacuums, steam-cleaning units, automatic cage-washers, and many types of laboratory equipment. Explosive materials are not commonly used, however, crystallized picric acid and previously opened and expired cans of ether are common potential explosion hazards. Class D materials are not common in animal care facilities but might exist in some laboratories.

Class B liquids are classified according to their flash point, the lowest temperature at which a liquid will produce vapor sufficient to propagate a flame. Flammable liquids have flash points less than 100°F. Combustible liquids have flash points greater than 100°F but less than 200°F. The flash points of combustible liquids are higher, so they are more difficult than are flammable liquids to ignite at room temperature. Knowledge of flash points of materials can be helpful in selecting a less-flammable material for a particular use so as to lower the related fire hazard. Material Safety Data Sheets for chemicals include information on flash points. (See page 42, Chemical Hazards.) OSHA provides very strict regulations for the storage and use of flammable and combustible liquids (29 CFR 1910.106).

Pressure Vessels

Compressed-gas cylinders, air receivers, high-pressure washing equipment, hydraulic lift lines, and steam generators house high-pressure air lines (over 30 psi), and autoclaves contain steam and contents under high pressure. These vessels present a substantial hazard to workers if uncontrolled or improper release of the pressure occurs. Compressed-gas cylinders should be secured at all times.

Lighting

One characteristic of animal care facilities that is not seen in many other operations is a fixed light-dark cycle. In animal care rooms, light cycles can vary, and most animals receive only artificial light. Animals can be kept on light-dark cycles that do not match the natural daily cycles. Or animals might be kept in rooms with single-color lights (usually red) or very low light. For humans, poor lighting can cause visual fatigue or create safety hazards that cause trips, slips, or falls. They might bump into corners of cages or other objects because they cannot see them easily in low light. Humans need an adjustment period for their eyes to become accustomed to the color or light levels in the room. Waiting for this adjustment will make work in the room easier and safer.

Electricity

Electric hazards can be present whenever electric current is flowing. Electric hazards are ubiquitous in animal care. Most of the hazards are obvious, such as the absence of a plate on a wall socket, an open electric panel, or an ungrounded plug. Less obvious hazards are present on cage-changing tables, biological safety cabinets, and wet vacuum systems. The electric hazards associated with those and other kinds of equipment can be minimized or eliminated through such engineering controls as ground-fault interrupters, such operational procedures as the use of lockout and tagout procedures to control energy sources during repair and maintenance of equipment (CFR 1919.147), and vigilance. Equipment that has frayed or exposed wires or that is designed to be connected to an ungrounded receptacle (as with a two-pronged plug) should not be used.

Ultraviolet Radiation

Exposure to ultraviolet (UV) radiation can occur in some operations involved in the care and use of laboratory animals. For example, UV germicidal lamps are used to sterilize clean surfaces in some work areas, and UV radiation is used in sterilizing water and in the diagnosis of fungal diseases. The most important exposures to UV radiation might be those of workers who perform outside work. UV radiation is divided into three classes designated UV-A, UV-B, and

TABLE 3-1 Classification and Description of Ultraviolet Radiation

UV Classification	Wavelengths (nm)	Effects	Sources
UV–A (black-light region)	320–400	Pigmentation of skin	Sunlight, black light
UV–B (erythemal region)	280–320	Photokeratitis, cataracts, erythema	Sunlight, artificial sources
UV–C (germicidal region)	100-280	Germicidal effects	Germicidal lamps

Sources: Adapted from National Safety Council 1988, pp. 227–232, and from the American Conference of Governmental Industrial Hygienists (ACGIH), 1994, p. 100.

UV-C, whose wavelengths, effects, and sources are shown in Table 3-1. UV radiation reacts with the vapors of chlorinated solvents—such as trichloroethylene, trichloroethane, and chlorofluorocarbons—to produce phosgene, a potent lung irritant. Those solvents should not be used in areas where UV-B or UV-C radiation is present.

If employees must work in the presence of UV radiation, their eyes and skin should be protected against UV exposure. Interlocking devices can be used to turn off UV sources before exposed areas are entered. Window glass is very effective at filtering out wavelengths less than 320 nm except for very intense sources.

Lasers

Laser is an acronym for **l**ight **a**mplification by the **s**timulated **e**mission of **r**adiation. Laser emissions are produced by solid-state, gaseous, and semiconductor lasers. Most states require lasers to be registered. The American National Standards Institute (ANSI Z–136.1 1986) has classified lasers on the basis of their power level and hazard potential as follows:

• *Class I*. Lasers that under normal operating conditions do not emit a hazardous level of radiation.
• *Class II*. Low-power lasers that do not have enough power to injure someone accidentally but do have enough power to cause injury if the beam is viewed for extended periods.
• *Class III*. Class IIIa, higher-power lasers that can cause injury if the beam is concentrated with a viewing device, such as binoculars; Class IIIb, lasers that can produce injury if viewed directly. The beam reflected off a mirror-like surface is also hazardous.

- *Class IV.* Lasers that, in addition to the conditions in Class III, can present a fire hazard.

The major hazard associated with lasers is related to the beam. The beam can cause burns, eye damage, lacerations, or fires, depending on its power. In animal care operations, lasers might be used to perform surgery or to provide medical treatment. Personnel who work with or around lasers should be trained in the hazards and the means to protect themselves. In the case of higher-power lasers, enclosing or shielding the beam (if possible) and providing interlocks on doors where a laser will be used are effective ways to reduce exposure to the beam. Laser surgery can also produce substantial aerosols, fumes, and toxic gases. These hazards should be controlled to prevent harmful exposures of employees.

All lasers use electric power, some in large quantities, so the risk of electric shock should be considered and reduced. The National Safety Council (NSC 1988) has produced a list of possible steps for reducing the risk of electric shock associated with lasers.

Ionizing Radiation

Ionizing radiation is ubiquitous in our daily lives. We are exposed to cosmic radiation, radon gas, natural background radiation, medical x rays, and even internal radiation from potassium-40. To be classified as ionizing, radiation must have enough energy to remove electrons from atoms and so create ions. The ionization can cause chemical changes that can be harmful to a living organism. Ionizing radiation can be classified as particulate and nonparticulate. Particulate radiation is composed of particles that are of atomic origin. Alpha particles are charged particles that each contain two neutrons and two protons. Beta particles are electrons that are emitted with very high energy from many radioisotopes. Positively charged counterparts of beta particles are called positrons. Alpha particles do not travel more than 0.5 in (1.3 cm) in air and cannot penetrate the dead layer of skin. The distance that beta particles can travel depends on their source: in air, some of the more energetic beta particles, such as those from phosphorus-32, can travel up to 30 ft (9 m), but beta particles from tritium (hydrogen-3) travel only 0.02 ft (0.6 cm). Beta particles are usually stopped by the skin but can cause serious damage to skin and eyes.

Nonparticulate radiation includes x rays and gamma rays. X rays and gamma rays are electromagnetic radiation with very short wavelengths. They are photons of energy and can penetrate matter. Photons are relatively difficult to shield. Gamma rays arise from nuclear decay; x rays arise from electron dislocation. When a radionuclide decays, it might produce alpha particles, gamma rays, beta particles, neutrons, or combinations of these. Irradiators and diagnostic x-ray machines are commonly used in research settings. Appropriate training of personnel and personal protection should be provided. Preventive maintenance of equipment is also critical to safe operations.

Radiation can present a hazard through inhalation, ingestion, skin contact, or proximity. The biological effect of ionizing radiation depends on the type of radiation, its energy, and the type of tissue that absorbs it. Two types of hazard must be considered: external and internal. A radionuclide that presents a radiation hazard when it is outside the body constitutes an external hazard; a radionuclide that presents a radiation hazard when it is ingested, inhaled, or absorbed constitutes an internal hazard. Alpha and beta particles do not travel very far in air, so they present mainly internal hazards; they can produce harm by being near tissue. Some of the more-energetic beta particles can present an external hazard.

Experimentation involving animals and radioisotopes is common in molecular biology today. Use of radioisotopes in or with animals presents several new hazards that must be dealt with. For example, some isotopes can be concentrated in a specific organ, such as iodine in the thyroid. Tissue that has concentrated a radioactive material might have to be handled or disposed of differently, depending on the isotope and the concentration. Bedding material from experimental animals exposed to radioactive materials should be surveyed to determine its radioactivity and then disposed of according to applicable regulations. If an isotope could be released by exhalation, additional engineering controls might be required. The use of radioisotopes is strictly controlled by the US Nuclear Regulatory Commission (US Congress 1971). Investigators should be authorized to use radioisotopes by their institutions; authorizations are based in part on evidence of training and established work practices.

Housekeeping

Good housekeeping keeps work surfaces clean and clear of obstructions, waste, and other material. If boxes, hoses, or bags of bedding material are not removed from the work area, trip hazards can be created or safe work might be impossible because working conditions are cramped. The act of cleaning itself sometimes creates hazards. For example, during steam cleaning of walls and floors of an animal room, the hoses can cause tripping hazards, high-temperature steam can cause burns, and wet floors can cause slipping hazards. Material left in hallways that are used for emergency egress poses a very serious hazard. Immediate removal of blockages of exits is imperative. Poor housekeeping practices can increase the seriousness of other hazards associated with animal care. For example, sweeping bedding, hair, and dander from floors, rather than using a vacuum cleaner with a filtered exhaust, can result in high concentrations of airborne allergens that can be distributed throughout the animal facility.

Ergonomic Hazards

Physical trauma can occur when workers perform tasks that require repetitive motions and lifting of heavy loads. Injuries that result from repetitive small stresses are often termed cumulative injuries. Cumulative injuries are not associated with a specific exposure incident. Common cumulative injuries include back injuries, carpal-tunnel syndrome, tennis elbow, and bursitis. Activities in animal care operations that contribute to back injuries include lifting heavy bags of feed, lifting heavy animals, lifting small weights incorrectly, moving or lifting cages, or clipping animals' fur manually. Adjusting control knobs, using a screwdriver, using pliers, opening and closing cage doors, moving small animals from cage to cage, operating video display terminals for extended periods, and mopping floors can also lead to repetitive-stress injuries. To reduce hazards due to repetitive motion, vary tasks to lessen the number of repetitions, re-engineer tasks, or redesign equipment or tools to require fewer repetitions with less strain.

Lifting heavy loads that exceed permissible-load recommendations of the National Institute for Occupational Safety and Health (NIOSH 1991) is unsafe and presents a substantial risk of acute injury. Anyone lifting heavy loads should be physically fit, should avoid sudden movements, and should use a two-handed lifting technique. Animal care operations that involve a potential for substantial physical stress include moving and restraining large animals, lifting and moving cages, lifting large feed bags, and moving high-pressure wet-vacuum systems. Engineering controls—such as the use of lifting equipment, automation of the lifting operation, or splitting of the load—can reduce the risk.

Once a hazard is recognized, employee education and engineering controls can be applied to reduce the potential for these types of injuries. Training should be updated if new tools are used in an operation and updated periodically to remind employees of proper work techniques. Employee involvement should be part of each solution.

Machinery

Conveyor belts, sanders, floor polishers, cage washers, room washing equipment, and other machinery have potential to cause injury. The common types of hazards presented by machinery are in-running nip points, crush points, and pinch points. In-running nip points are places on a roller or similar moving surface where a body part of an exposed worker could be pulled into the machinery. Crush points and pinch points are areas of a machine where two surfaces could come together to crush or pinch part of the body. These all occur in machinery that has exposed moving parts. Each machine should be evaluated to determine whether a worker's hand or arm could be placed in an area where it could be injured. If a hazardous area is identified, guarding should be installed to eliminate the hazard. Guarding is important even when workers know that they are not to place their hands in a dangerous area. Slips, falls, and loss of awareness of the hazard can cause injury if guarding is not in place. In large equipment that requires operators or repair mechanics to work in the operating chamber, such as cage washers, an internal release mechanism should be available to allow emergency escape if the equipment is inadvertently started.

Noise

Exposure to intense noise can result in loss of hearing. Chronic noise-induced hearing loss is a permanent condition and cannot be treated medically. This type of hearing loss is usually characterized by declining sensitivity to frequencies above 2,000 Hz. Exposure to an intense noise for a short period can cause temporary or permanent loss of hearing. OSHA limits employee exposure to noise to 90 decibels measured on the A scale of a standard sound-level meter at slow response (dBA) averaged over an 8-h workshift (29 CFR 1910.95). The time-weighted average must be lower than 90 dBA if the workshift is longer than 8 h (29 CFR 1910.95). Where levels exceed 85dBA, the exposed employees need to participate in a hearing-conservation program that includes monitoring, audiometric testing, hearing protection, training, and record-keeping (29 CFR 1910.95 c though o). Hearing loss is not the only adverse effect of exposure to noise. Noise can make speech difficult, cause loss of concentration, distract workers, and increase fatigue (NSC 1988).

In an animal care facility, noise can result from animals, particularly pigs and dogs, and from equipment, such as cage washers, high pressure air cleaning equipment, and wet vacuum systems operated in a confined space. A useful way of assessing whether a noise exposure might be excessive is to visit the area and attempt to converse with another person or attempt to talk on the telephone. The noise is probably excessive if normal speech or talking on the telephone is difficult or impossible. When this condition is observed, the noise levels should be assessed by a person knowledgeable about noise, noise-measurement techniques, and data interpretation. Most often, such a person will be an industrial hygienist or an acoustical engineer. OSHA requires that engineering controls be applied first to control the hazard. Engineering controls include shielding, quieter equipment, and installation of sound-deadening materials on the walls and ceilings. If acceptable noise levels are not achieved that way, administrative controls or personal protective equipment will be necessary. Administrative controls include limiting the time that an employee works in the noise-hazard area. It is prudent to provide workers who are exposed to a noise hazard earplugs, earmuffs, or other protective equipment during the noise-evaluation period.

Ultrasonography is used in laboratories and animal care facilities for imaging internal structures. If the frequency is below 20 kHz, it is covered by the OSHA noise standard. Even if it is above 20 kHz, noise exposure is possible because of subharmonics at these higher frequencies (Strickoff and Walters 1990).

Chemical Hazards

Most employees engaged in the care and use of research animals are familiar with the hazards of chemicals used in animal care and laboratory environments. Employee knowledge of chemical hazards and of relevant protective measures has been focused and increased in recent years through employers' responses to two important health and safety standards promulgated by OSHA: the Hazard Communication Standard (29 CFR 1910.1200) and the Occupational Exposure to Hazardous Chemicals in Laboratories (29 CFR 1910.1450), which is known as the laboratory standard. The recognition and control of chemical hazards in research institutions have also been aided by *Prudent Practices for Handling Hazardous Chemicals in Laboratories* (NRC 1981). That volume was extensively revised and updated in 1995, and the new edition, *Prudent Practices in the Laboratory: Handling and Disposal of Chemicals* (NRC 1995), provides practical guidance for evaluating chemical hazards and for working safely with chemicals in the research setting. It extensively discusses sources of hazard information and principles for evaluating and elucidating toxic effects of chemicals. It constitutes a relevant and comprehensive reference document on the recognition and control of chemical hazards, and it should be consulted by all who have responsibility for the planning, conduct, and support of safe research.

Flammability, corrosiveness, reactivity, and explosivity are hazardous properties of chemicals that are usually well understood. Toxicity is the least-predictable hazardous property of chemicals. Exposure to toxic chemicals can cause acute or chronic health effects. General classes of toxic chemicals that might be handled in a research environment are carcinogens, allergens, asphyxiants, corrosives, hepatotoxicants, irritants, mutagens, nephrotoxicants, neurotoxicants, and teratogens. Health risks associated with toxicants depend on both the inherent toxicity of the chemicals and the nature and extent of exposure to them. Animal care activities can seriously influence the potential for employee exposure. Thus, animal care practices that might contribute to employee exposures need to be carefully assessed so that toxic hazards of chemicals associated with the care and use of research animals can be recognized and controlled. A comprehensive review of chemical-hazard assessment and control is provided in *Prudent Practices in the Laboratory: Handling and Disposal of Chemicals* (NRC 1995).

Typical sources of chemical exposure in the care and use of research animals involve the use of disinfectants, pesticides, anesthetic gases, and chemicals for preserving tissues. Sources can include animals that have been intentionally exposed to highly toxic chemicals. Another important source is the disposal of bedding and other waste materials from experimental procedures.

Disinfectants and detergents include soaps, cleaning chemicals, acid-containing chemicals, alcohols (most commonly ethanol and isopropanol), aldehydes (including formaldehyde and gluteraldehyde), and halogenated materials (such as chlorinated and iodinated bleaches). Some phenolic compounds (including potassium *o*-phenylphenate and potassium *o*-benzyl-*p*-chlorophenate) and quaternary ammonium compounds are also used as disinfectants. Various pesticides can be used within animal facilities, but most animal facilities restrict the use of pesticides because of their potential effects on the animals. The primary chemical used as a preservative is formalin as a 10% neutral-buffered solution, but other materials are used from time to time.

Several occupational diseases—including cancer, spontaneous abortion, and liver disease—have been associated with exposure to waste anesthetic gases. Monitoring exposures to waste anesthetic gases in animal operating rooms is an important part of the health and safety program because of the difficulty in matching anesthetic-delivery equipment to the animals.

Burns and irritation of the skin are the most common chemical injuries associated with animal care and use. Some chemicals, such as formaldehyde and gluteraldehyde used for preserving tissue, can cause an allergic response in sensitized people. The risk of injury and illness associated with chemical use can be minimized by practices that reduce or prevent exposure.

HAZARDS ASSOCIATED WITH EXPERIMENTAL PROTOCOLS

A fundamental principle in the conduct of research is the need to determine the potential hazards associated with an experiment before beginning it. That is extremely important in planning experiments that involve research animals, because investigators might be unfamiliar with the intrinsic hazards presented by the animal species of choice or tissues derived from them, and managers and their employees who care for the research animals should be informed of the hazards presented by the experimental protocol. Consideration of both animal-related hazards and protocol-related hazards would benefit from a collaborative assessment in which the investigator, the institutional veterinarian, the animal care supervisor, and a health and safety professional participate. A collaborative assessment is strongly encouraged if the animal experimentation involves either the testing of chemicals for their toxic properties or research with experimentally or naturally infected animals. Whether or not a collaborative initiative is pursued, investigators have an obligation to identify hazards associated with their research and to select the safeguards that are necessary to protect employees involved in the care and use of their research animals.

Hazards associated with experimental protocols are influenced by two principal factors: the dangerous qualities of the experimental agents and the complexity or type of the experimental operations. For example, toxicity, reactivity, flammability, and explosivity should be considered when an experimental protocol involving chemical agents is being planned, and virulence, pathogenicity, and communicability are possible hazardous qualities of biological agents.

The complexity and type of an experimental operation have a direct impact on the extent of potential exposure that an employee receives while carrying out or participating in an experimental protocol. For example, during incorporation of a test chemical into feed for ingestion studies, a contaminated dust created during milling and mixing and during transfer of the diet could result in respiratory and dermal exposures. Test material applied to the skin of experimental animals might be disseminated by handling of animals, clipping of hair, changing of bedding, and sweeping of the animal room floor. Vapors are potential sources of exposure during the application of test material to the skin. Exposing an animal to an agent by injection will create a risk of accidental self-inoculation. Inhalation challenges are particularly hazardous and should be conducted only by investigators who have appropriate experience and containment equipment.

Protocols Involving Chemicals of Unknown Hazard

A comprehensive, rigidly followed plan is necessary for testing chemicals of unknown hazard for their toxic properties. It should be presumed that a chemical is hazardous to humans, and the plan should describe specific procedures for handling the chemical from receipt through disposal of animal waste and processing of tissues for histopathological or biochemical examination. *Prudent Practices in the Laboratory: Handling and Disposal of Chemicals* (NRC 1995) provides an excellent general model for planning experiments that involve hazardous chemicals. It was specifically structured to follow the sequence of stages that should be considered in planning a safe experiment: evaluating hazards and assessing risks in the laboratory, management of chemicals, working with chemicals, working with equipment, disposal of chemicals, laboratory facilities, and government regulation of laboratories. It is important not to underestimate the risk presented by experimental chemicals. But most references on chemical safety provide little guidance that is directly applicable to the care and use of research animals. Therefore, developing plans for a specific research protocol that involves research animals and chemicals of unknown hazard will require ingenuity, a quality best derived from a collaborative planning process.

Protocols Involving Infectious Agents

Experiments involving experimentally or naturally infected research animals present recognized risks of occupationally acquired infections. In the largest survey of laboratory-acquired infections conducted to date, research animals or their ectoparasites were associated with about 17% of the reported infections (Pike 1976). In the few cases (under 3%) in which infections were attributed to a recognized accident, the primary source was a bite or scratch from an infected animal. That survey and others (Sullivan and others 1978) have shown that trained scientific personnel and technicians were most likely to be infected, although animal care providers and janitorial and maintenance workers have been proved to be at risk for occupationally acquired infection. Most of the zoonotic infections cited in these surveys were associated with research activities involving experimentally infected animals. Transmission of zoonotic disease in an animal facility that is not involved with infectious disease research is rare. CDC and NIH have identified 17 infectious agents or genera other than arboviruses as proven hazards for personnel who use and care for experimentally or naturally infected research animals (CDC-NIH 1993). The agents and genera are summarized in Table 3-2. Arboviruses—most notably Venezuelan equine encephalomyelitis virus, yellow fever virus, Rift Valley virus, and Chikungunya virus—have also been responsible for laboratory animal-associated infections (Hanson and others 1950). The Subcommittee on Arbovirus Laboratory Safety (SALS) of the American Committee on Arthropod-Borne Viruses reported 818 occupationally acquired infections caused by 62 different arboviruses or related viruses (SALS 1980). A total of 19 of these infections, which were associated with 10 viruses—Semliki Forest, Venezuelan equine encephalitis, Western equine encephalitis, yellow fever, Hypr, Rift Valley fever, Congo-Crimean hemorrhagic fever, Junin, Lassa, and Machupo—resulted in death.

Investigators who are planning research activities involving experimentally or naturally infected vertebrate animals should carefully review *Biosafety in Microbiological and Biomedical Laboratories* (CDC-NIH 1993). It defines four levels of control that are appropriate for animal research with infectious agents that present occupational risks ranging from no risk of disease for healthy people to high individual risk of life-threatening disease, and it recommends guidelines for specific agents. The four levels of control, referred to as animal biosafety levels 1-4, each have appropriate microbiological practices, safety equipment, and features of animal facilities. The selection of an animal biosafety level is influenced by several characteristics of the infectious agent, the most important of which are the severity of the disease, the documented mode of transmission of the infectious agent, the availability of protective immunization or effective therapy, and the relative risk of exposure created by manipulation in handling the agent and caring for infected animals.

Animal biosafety level 1 is the basic level of protection appropriate for well-characterized agents that are not known to cause disease in healthy humans. Animal biosafety level 2 is appropriate for handling a broad spectrum of moderate-risk agents that cause human disease by ingestion or through percutaneous or mucous-membrane exposure. Extreme precautions with needles or sharp instruments are emphasized at this level. Animal biosafety level 3 is appropriate for agents that present risks of respiratory transmission and that can cause serious and potentially lethal infections. Emphasis is placed on the control of aerosols by containing all manipulations and housing infected animals in isolators or ventilated cages. At this level, the animal facility is designed to control access to areas where animals are kept and includes a specialized ventilation system that is designed to maintain directional airflow. Exotic agents that pose a high individual risk of life-threatening disease by the aerosol route and for which no

TABLE 3-2 Reported Occupationally Acquired Infections Associated with Experimentally or Naturally Infected Research Animals

Pathogenic Agent	Animals	Comment	References
Viral Agents			
B virus (*Circopithecine herpesvirus 1*) (formerly *Herpesvirus simiae*)	Macaques	Contact with experimentally and naturally infected animals	Holmes and others 1990, Palmer 1987
Hepatitis A virus	Nonhuman primates	Contact with experimentally and naturally infected animals	Pike 1979
Lymphocytic choriomeningitis virus	Mice, hamsters, guinea pigs	Contact with experimentally and naturally infected animals	Bowen and others 1975, Jahrling and Peters 1992, Pike 1976
Marburg virus	African Green monkeys	Contact with naturally infected animals	Martini & Siegert 1971 Martini 1973
Simian immunodeficiency virus	Macaques	Handling of blood from experimentally infected animals	CDC 1992a, Khabbaz and others 1992
Vesicular stomatitis virus	Livestock	Contact with naturally infected animals	Hanson and others 1950, Patterson and others 1958
Rickettsial Agents			
Coxiella burnetii	Sheep	Contact with naturally infected animals	CDC 1979, Spinelli and others 1981
Bacterial Agents			
Brucella (*B. abortus, B. canis, B. melitensis, B. suis*)	Cattle, dogs, goats, swine	Contact with experimentally and naturally infected animals, presumed aerosol exposure	Pike 1976
Campylobacter jejuni	Dogs, primates, coyotes, etc.		Fox and others 1989
Chlamydia psittaci	Birds	Contact with experimentally and naturally infected animals, presumed aerosol exposure	Miller and others 1987, Pike 1976

TABLE 3-2 Continued

Pathogenic Agent	Animals	Comment	References
Francisella tularensis	Rabbits	Contact with experimentally and naturally infected animals or their ectoparasites	Pike 1976
Leptospira interrogans	Rabbits, dogs, rats, mice, guinea pigs	Contact with experimentally and naturally infected animals	Richardson 1973
Legionella pneumophila	Guinea pigs	Aerosol or droplet exposure during animal challenge	CDC 1976
Mycobacterium tuberculosis	Nonhuman primates	Contact with experimentally and naturally infected animals	Kaufmann and Anderson 1978
Salmonella spp.	Mice, rats, dogs, cats	Contact with experimentally and naturally infected animals	Grist and Emslie 1987, Miller and others 1987, Pike 1976
Shigella spp.	Guinea pigs, rats, mice, nonhuman primates	Contact with experimentally infected animals	Pike 1976
Streptobacillus moniliformis	Rats	Contact with experimentally and naturally infected animals	Pike 1976
Fungal Agents			
Sporothrix schenckii	Rats	Bite from an experimentally infected animal	Jeanselme and Chevallier 1910, 1911
Microsporum, Trichophyton	Mice, rabbits, guinea pigs	Contact with experimentally and naturally infected animals	Hanel and Kruse 1967, McAleer 1980; Pike 1976

Source: CDC-NIH 1993.

treatment is available are restricted to animal biosafety level 4 high-containment facilities. Worker protection in these facilities is provided by the use of physically sealed glove boxes or fully enclosed barrier suits that supply breathing air. Most research involving experimentally and naturally infected vertebrate animals will be conducted at animal biosafety level 2 or 3. A summary of hazard control elements for these two animal biosafety levels is presented in Table 3-3. Animal biosafety level 1 is not addressed here because it represents normal housing without special precautions. Animal biosafety level 4 is not discussed because containment facilities for this work are limited to a few highly specialized institutions that have considerable experience in the handling of dangerous and exotic pathogens.

Research protocols involving emerging and re-emerging pathogens require careful planning and might require review of previous studies. Most of the literature on safety in handling infectious agents was published 3-4 decades ago, but it is still invaluable in planning safe experiments. Modern research can also present novel hazards that require careful review. For example, the potential occupational health and safety risks need to be considered before animal experiments are undertaken to evaluate the safety to humans of viral vectors that are being proposed for use in gene therapies. Similarly, studies with transgenic animals that express receptors for human pathogens or whose genomes contain proviral DNA for an infectious virus should be evaluated to determine whether safeguards appropriate for handling the wild-type infectious agent should be applied. Assistance in the determination of risk and the selection of appropriate safeguards can be found in the NIH *Guidelines for Research Involving Recombinant DNA Molecules* (NIH 1994). A recent revision of its Appendix B has a section (B-V) on frequently used viral agents, including viral vectors. These protocols might require approval of the institution or funding agency. The institution's biosafety committee and biosafety officer are valuable resources and should be consulted when experiments are being planned.

Several authoritative reference works provide excellent guidance for the safe handling of infectious microorganisms in research. Three that are particularly noteworthy are *Biosafety in the Laboratory: Prudent Practices for the Handling and Disposal of Infectious Materials* (NRC 1989), *Laboratory Safety: Principles and Practices* (Fleming and others 1995), and *Biosafety in Microbiological and Biomedical Laboratories* (CDC-NIH 1993).

Most-helpful practices to prevent occupationally acquired infections associated with the care and use of research animals are the following:

- Avoid the use of sharps whenever possible. Take extreme care when using a needle and syringe for inoculating research animals or when using sharps during necropsy procedures.
- Keep hands away from mouth, nose, and eyes.

TABLE 3-3 Summary of Recommended Animal Biosafety Levels 2 and 3 for Research Programs that Involve Experimentally or Naturally Infected Vertebrate Animals

Animal Biosafety Level	Agents	Practices	Safety Equipment	Facilities
2	Cause human disease of varied severity; indigenous Hazard: percutaneous exposure, mucous membrane exposure, ingestion (enteric pathogens) Examples: HBV, HIV, *Shigella flexneri*, *Salmonella typhimurium*, *Toxoplasma gondii*	• Good personal hygiene • Limited access • Biohazard warning signs • Sharps precautions • Biosafety manual • Personal protective • Decontamination of cages before washing • Decontamination of infectious waste	• Primary barriers: containment equipment used for necropsy and procedures with high potential for creating aerosols equipment: laboratory coats, gloves, face and respiratory protection as required	• No recirculation of exhaust air • Autoclave available • Hand washing sink in animal rooms
3	Cause human disease with serious or lethal consequences; indigenous or exotic Hazard: aerosol transmission Examples: *Mycobacterium tuberculosis*, *Brucella canis*, *Coxiella Burnetii*	• Good personal hygiene • Controlled access • Biohazard warning signs • Sharps precautions • Biosafety manual • Decontamination of clothing before laundering • Decontamination of cages before bedding is removed • Decontamination of all wastes	• Primary barriers: containment equipment used for all activities involving infectious materials or infected animals • Personal protective equipment: laboratory gowns, gloves, face and respiratory protection as required, protective footwear as required	• Physical separation from access corridors • Self-closing double-door passage way • No recirculation of exhaust air • Directional airflow • Hand washing sink in animal rooms • Autoclave available in facility

Source: CDC-NIH 1993.

- Wear protective gloves and a laboratory coat or gown in areas where research animals are kept.
- Remove gloves and wash hands after handling animals or tissues derived from them and before leaving areas where animals are kept.
- Use mechanical pipetting devices.
- Never eat, drink, smoke, handle contact lenses, apply cosmetics, or take or apply medicine in areas where research animals are kept.
- Perform procedures carefully to reduce the possibility of creating splashes or aerosols.
- Contain operations that generate hazardous aerosols in biological safety cabinets or other ventilated enclosures.
- Wear eye protection.
- Keep doors closed to rooms where research animals are kept.
- Promptly decontaminate work surfaces after spills of viable materials and when procedures are completed.
- Decontaminate infectious waste before disposal.
- Use secondary leakproof containers to store or transfer cultures, tissues, or specimens of body fluids.

4

Allergens

Allergic reactions to animals are among the most common conditions that adversely affect the health of workers involved in the care and use of animals in research. One survey (Lutsky 1987) demonstrated that three-fourths of all institutions with laboratory animals had animal-care workers with allergic symptoms. The estimated prevalence of allergic symptoms in the general population of regularly exposed animal-care workers ranges from 10% to 44% (Hollander and others 1996, Knysak 1989). An estimated 10% of laboratory workers eventually develop occupation-related asthma.

Attempts have been made to determine whether persons with allergic conditions, such as allergic rhinitis (hay fever), are at higher risk than normal persons of developing animal-dander sensitivity when working with laboratory animals. On the basis of current estimates, up to 73% of persons with pre-existing allergic disease eventually develop allergy to laboratory animals (Agrup and others 1986, Platts-Mills and others 1986, Venables and others 1988). Allergy is most often manifested by nasal symptoms, itchy eyes, and rashes. Symptoms usually evolve over a period of exposure of 1-2 years. Occupation-related asthma, a more serious disorder, might develop in about 10% of persons with allergic disease who work with laboratory animals (Hunskaar and Fosse 1993). Occupation-related asthma not only can cause symptoms of cough, wheezing, and shortness of breath while the worker is exposed to laboratory animals, but also can lead to chronic symptoms (persisting for months to years) even after exposure ceases.

Workers exposed to laboratory animals can be categorized into several risk groups. The information cited above is shown in Table 4-1 for four risk groups based on history of allergic disease and sensitization to animal proteins. Except in

TABLE 4-1 Risk of Developing Allergy to Laboratory Animals

Risk Group	History	Risk of allergic reactions to laboratory animals	Comment
Normal	No evidence of allergic disease	~10%	90% of normal group will never develop symptoms in spite of repeated animal contact
Atopic	Pre-existing allergic disease	Up to 73%	Workers who become sensitized to animal proteins will eventually develop symptoms on exposure
Asymptomatic	Immunoglobulin E antibodies to allergenic animal proteins	Up to 100%	Risk of developing allergic symptoms of rhinitis, asthma, or contact urticaria with continued exposure is high
Symptomatic	Clinical symptoms on exposure to allergenic animal proteins	100%	33% with chest symptoms; 10% of group might develop occupational asthma; even minimal exposure can lead to permanent impairment

a few situations, a dose-response relationship that defines sensitization, induction of disease, and production of symptoms in association with specific allergen concentrations has not been established.

Contact urticaria ("hives") is typically due to the application of an allergen (usually a protein or glycoprotein) directly onto the skin. A common example is the development of wheal and flare reactions that produce welts when a person's skin and the tail of a mouse or rat come into contact. Scratches by cats and dogs can produce similar responses. Latex in rubber gloves is another cause of contact urticaria.

Although symptoms of asthma in laboratory-animal workers are most obvious in the work environment, they can also occur at night and awaken sufferers. In almost all asthmatic people with laboratory-animal allergy, nasal and eye symptoms preceded the development of asthma (Bland and others 1987).

In rare instances, a person who has become sensitized to an animal protein in the saliva of the animal experiences a generalized allergic reaction termed anaphylaxis when bitten by an animal (Teasdale and others 1993). People working in entomology laboratories can be exposed to stinging insects, such as bees, wasps and ants, which can cause similar reactions. Anaphylaxis can be evident as diffuse itching, hives, and swelling of the face, lips, and tongue. Some people experience difficulty in breathing because of laryngeal edema; others develop asthma with wheezing. In some instances, shock can lead to loss of consciousness. Anaphylactic reactions vary from mild generalized urticarial reactions to profound life-threatening reactions.

MECHANISMS OF ALLERGIC REACTIONS

The allergic reactions described above are examples of classic immunoglobulin E-mediated reactions. Such reactions are the consequence of a series of immunological and biochemical events. First, a person is exposed to the allergen, which is usually a protein or glycoprotein. In the case of laboratory animal allergy, the route of exposure is most often due to airborne allergens (see Table 4-2). The allergen is processed by the macrophages or B lymphocytes and presented to T lymphocytes. Helper T lymphocytes stimulate B lymphocytes to produce antibodies of the immunoglobulin E (IgE) class specific for the allergen.

TABLE 4-2 Allergic Reactions to Laboratory-Animal Allergens

Disorder	Symptoms	Signs
Contact urticaria	Redness, itchiness of skin, welts, hives	Raised, circumscribed erythematous lesions
Allergic conjunctivitis	Sneezing, itchiness, clear nasal drainage, nasal congestion	Conjunctival vascular engorgement, cheminosis, clear discharge (usually bilateral)
Allergic rhinitis	Sneezing, itchiness, clear nasal drainage, nasal congestion	Pale or edematous nasal mucosa, clear rhinorrhea
Asthma	Cough, wheezing, chest tightness, shortness of breath	Decreased breath sounds, prolonged expiratory phase or wheezing, reversible airflow obstruction, airway hyperresponsiveness
Anaphylaxis	Generalized itching, hives, throat tightness, eye or lip swelling, difficulty in swallowing, hoarseness, shortness of breath, dizziness, fainting, nausea, vomiting, abdominal cramps, diarrhea	Flushing, urticaria, angioedema, stridor, wheezing, hypotension

IgE is found in the circulation in low concentrations and binds to mast cells and basophils. Mast cells are abundant in the respiratory tract, gastrointestinal tract, and skin, the main sites of allergic reactions. When a person so "sensitized" is re-exposed to the same allergen, the allergen binds to IgE molecules and causes the release of histamine and other chemical mediators stored in the mast cells and basophils. The mediators, on contact with the relevant tissues, can produce hives, nasal congestion, sneezing, nasal drainage, coughing, wheezing, and shortness of breath.

All those reactions are termed "immediate hypersensitivity" responses because they are noted within 10-15 min of exposure to the allergen. However, it is now recognized that such reactions not only can occur immediately but also have a late component; that is, the symptoms can recur 4-6 h after exposure without further allergen stimulation.

Virtually all human beings are capable of developing allergic reactions; however, some individuals are more susceptible. These people (atopics) are more likely to develop IgE antibodies to allergens owing to an inherited tendency. This is an autosomal dominant trait with variable expression that has been linked to genetic markers on chromosome 5 (Blumenthal and Blumenthal 1996, Marsh and others 1994). Persons with atopy often develop allergic diseases, such as allergic rhinitis, asthma, and atopic dermatitis (eczema) when chronically exposed to allergens.

SPECIFIC ANIMALS THAT CAN PROVOKE ALLERGIC REACTIONS

Rats

Rats are among the most commonly used laboratory animals and are responsible for symptoms in a large portion of people who have laboratory-animal allergy. The major sources of rat-allergen exposure appear to be urine and saliva of the animal. A major rat-urine allergen with two isoforms has been identified: *Rat n 1A*, a pre-albumin, and *Rat n 1B* (α2-euglobulin) (Eggleston and others 1989, Longbottom 1980). These two proteins have some cross-reactivity, although they differ in molecular weight and isoelectric point. Their amino acid composition is similar, but their carbohydrate concentration differs. The amino acid sequence of *Rat n 1B* has been obtained (Laperche and others 1983).

Sampling methods have been developed to measure the amount of airborne allergen and the size of the airborne particles that contain rat allergen (Eggleston and others 1989; Platts-Mills and others 1986). Particles that contain rat allergen found in air samples from a rat vivarium vary from <0.5 to >20 µm in aerodynamic equivalent diameter. Disturbance of rat litter leaves a substantial proportion of the smaller particles airborne for 15-35 min (Platts-Mills and others 1986); most of these particles are easily respirable.

In a preliminary study of 335 workers exposed to rats, the risk of respiratory or skin symptoms was related to the duration of exposure to rat urinary protein concentrations of at least 1 µg/m^3 of air sampled (Tee and others 1993), the concentrations most likely to be encountered by animal-care technicians. Allergic symptoms due to exposure to rats were more likely to develop in atopic subjects (those with pre-existing sensitivity to nonanimal allergens) than in nonatopic subjects.

Exposure concentrations are clearly task-related. Cage-cleaning resulted in a mean airborne *Rat n 1* concentration of 21 ng/m^3 (range, 8.1-69 ng/m^3); handling rats for weighing, shaving, injections, and collection of blood and urine samples yielded a mean of 13 ng/m^3 (range, undetectable to 45 ng/m^3); and surgery on anesthetized animals or euthanasia of unconscious animals yielded a mean of 3 ng/m^3 (range, undetectable to 12 ng/m^3) (Eggleston and others 1989). It should be noted that these levels are an order of magnitude lower than reported by Tee and others (1993). This difference might be accounted for by the fact that Eggleston and co-workers measured for the specific allergen *Rat n 1*, whereas Tee and colleagues measured total airborne rat allergenic activity.

The importance of these exposures has been demonstrated in environmental challenge studies in which workers are exposed in rooms containing animals. Eggleston and co-workers (1990) measured airborne *Rat n 1* in a rat vivarium over the course of 1 h. The allergen concentration ranged from less than 1.5 to 310 ng/m^3 and was much higher during cage-cleaning than during quiet activity. Of 12 rat-allergic volunteers working in this environment, all had nasal symptoms and evidence of histamine release in their nasal secretions during the period of exposure, and five had decreases in pulmonary function greater than 10%. This experiment demonstrated that occupational exposure was directly correlated with the development of nasal symptoms and asthma in the sensitized volunteers.

Airborne allergen concentrations depend on the balance between the rate of allergen production and the rate of removal. And the magnitude of exposure to rat allergens is directly proportional to the number of animals in the area. Urine is a major source of allergen, and contact with contaminated litter seems to be a major source of exposure (Gordon and others 1992). Ventilation might be an effective means to lower exposure when production of allergen is low, because of either a small number of animals or little disturbance of litter, but it might be ineffective when production is high. For example, Swanson and others (1990) found that it might take up to 127 air changes per hour to reduce exposures sufficiently to make symptoms unlikely when many rats were present in the sampling area.

Mice

Mice are another important source of allergen exposure of laboratory workers. The major mouse allergen is a urinary protein, *Mus m 1*. *Mus m 1* has been molecularly cloned and its amino acid sequence deduced. It is analogous in many ways to *Rat n 1B* in that it is produced in the liver and saliva, is secreted in the urine, and has 80% amino acid sequence homology with *Rat n 1B* (Clark and others 1984). Urine samples contain *Mus m 1* at a concentration 100 times that in serum, and male mice excrete 4 times as much of it as female mice (Lorusso and others 1986).

Air-sampling techniques have been developed to monitor concentrations of major mouse urinary proteins in the environment (Twiggs and others 1982). Airborne allergen concentrations range from 1.8 to 825 ng/m^3, depending on the number of animals and the type of activity in the environment. The particles that contain most of the allergen vary from 6 to 18 µm in diameter (Price and Longbottom 1988). Sakaguchi and others (1989a) found that most of the airborne allergen in undisturbed air in a room containing 350 mice was trapped by a filter with a retention size greater than 7 µm. In disturbed air (in which cage-cleaning was conducted), allergen concentration increased by up to 5 times and the proportion of small particles (1.1 µm and smaller) increased by 3 times. Airborne concentrations are related to the number of mice present in the sampling area and the degree of work activity (Twiggs and others 1982).

Guinea Pigs

Immunochemical studies have identified allergenic components in the dander, fur, saliva, and urine of guinea pigs (Walls and others 1985); urine appears to be the major source of allergen. Most guinea pig allergen activity is associated with particles greater than 5 µm, but about 10% is found on particles smaller than 0.8 µm, which are small enough to penetrate into the lower respiratory tract (Swanson and others 1984).

Gerbils

Gerbils are occasionally used as laboratory animals, and allergic sensitivity to them has been reported (Gutman and Bush 1993). The allergens involved have not been identified.

Rabbits

Rabbits are used widely as laboratory animals and are a recognized cause of allergic symptoms in many workers. A major glycoprotein allergen has been described that appears to occur in the fur of the animals, and minor allergenic components found in rabbit saliva and urine have been identified (Warner and Longbottom 1991). Allergenic activity is associated with particles less than 2 µm in diameter (Price and Longbottom 1988).

Cats

Domestic cats are kept as pets by many people, and sensitization can occur outside the laboratory environment. Furthermore, allergy to cats might predispose workers to the development of allergy to laboratory animals, such as mice and rats (Hollander and others 1996). There is a close link between immunological sensitization and development of asthma in people sensitive to cats (Desjardins and others 1993). Those with pre-existing sensitivity might encounter worsening of their symptoms and possibly develop asthma during the course of their work exposure.

The major cat allergen is the protein *Fel d 1* (Kleine-Tebbe and others 1993). *Fel d 1* was first described by Ohman and colleagues (1974). It is produced in the sebaceous glands of the skin and coats the hair shafts (Woodfolk and others 1992). It is also produced in the saliva (Anderson and others 1985). *Fel d 1* has been molecularly cloned, its amino acid sequenced, and its allergenic structure analyzed (Morgenstern and others 1991). *Fel d 1* is found in all cats, and cross reactivity occurs throughout all species of cats. However, individual cats shed different amounts of the allergen (Wentz and others 1990), and male cats might shed more than female cats. A few people can become sensitized to cat albumin.

The size of particles that contain cat allergen varies, but many are less than 0.25 µm in diameter (Findlay and others 1983) and are easily carried deeply into the lung. Exposure to two cats that produced allergen at a concentration of 1.1-128 ng/m^3 was sufficient to cause symptoms of rhinitis and asthma in 10 persons with cat sensitivity (VanMetre and others 1986). Cumulative doses of 80-98 ng of *Fel d 1* inhaled over 2 min can cause a sufficient decrease in pulmonary function to produce an asthma attack. Placing one cat in a room with a volume of 33 m^3 increases the concentration of *Fel d 1* from nondetectable to 30-90 ng/m^3, which would be sufficient to cause an asthma attack within 25 min in a sensitized person (VanMetre and others 1986).

Airborne *Fel d 1* remains suspended for long periods because of its small particles (Luczynska and others 1990). The allergen appears to be highly electrostatically charged and therefore sticks to surfaces, such as walls and laboratory benches (Wood and others 1992). It can be transferred from those materials to hands, or the materials can act as reservoirs and can hold large quantities of allergen in the absence of cats.

Decreasing the airborne concentrations of cat allergen can be attempted by washing the animal (Middleton 1991; Ohman and others 1983). Using a filtered vacuum cleaner, removing carpeting, running a high-efficiency air cleaner, and washing the cat(s) can decrease concentrations in the air (deBlay and others 1991). Simply increasing ventilation rates from eight to 40 air changes per hour in a room containing two female cats did not reduce the clearance of airborne cat allergen (Wood and others 1993). After removal of cats from the environment, the time for concentrations to reach those seen in areas where there has been no cat can be 20 weeks or more (Wood and others 1989).

In addition to cats themselves, it is now recognized that cat fleas can produce allergic symptoms in some people (Baldo 1993).

Dogs

Like exposure to cats, exposure to domestic dogs outside the work environment can lead to sensitization and is also a risk factor for laboratory animal allergy (Hollander and others 1996). The major allergens of dogs are not as well studied as cat allergens, but an important allergen, *Can f 1*, has been identified (deGroot and others 1991; Schou and others 1991b). Collections of dust samples from homes with a dog in residence showed a *Can f 1* concentration of 120 µg/g of dust, compared with 3 µg/g where there was no dog (Schou and others 1991a). There is some question about cross reactivity among breeds of dogs, but the relevant information is not complete. Sources of exposure to dog allergens appear to be saliva, hair, and skin (Spitzauer and others 1993). Dog albumin has also been shown to be an important allergen (Spitzauer and others 1994). About 35% of people who are allergic to dogs have IgG antibody to albumin. The allergen has been molecularly cloned and shares amino acid sequence homology with other albumins.

Primates

Sensitization to primates is unusual. Despite widespread exposure to primates in research settings, few cases of sensitivity to primate allergens have been documented. Cases of sensitivity to lesser bushbaby (galago) and cottontop tamarin have been identified (Petry and others 1985). Allergenic activity was found in the dander of the latter. Whether other sources, such as saliva, are important is not clear.

Pigs

Asthma and other respiratory symptoms have been attributed to pig exposures, particularly in farm operations. In general, the symptoms do not appear to be allergic but more often are related to exposure to high nitrogen concentrations, especially in confinement operations (Matson and others 1983; Zhou and others 1991). Occupational asthma was described in a person who apparently had allergic sensitivity to a urinary protein from pigs (Harries and Cromwell 1982).

Cattle

Sensitivity to cattle has been reported in 15-20% of dairy farmers. The allergens have not been completely described, but components of dander and urine have been identified as allergenic (Ylönen and others 1992). A purified allergen with a molecular weight of 20-25 kilodaltons (kD) and an isoelectric point of 4.1 has been described (Ylönen and others 1994). Airborne cow-dander allergen concentrations in animal sheds range from 137 to 19,800 ng/m^3.

Horses

Horses constitute a highly potent source of allergens. The nature of the allergens has not been established, but a 27-kD allergen from horse dander, skin scrapings, and albumin are important (Fjeldsgaard and Paulsen 1993). They appear to be shed by the skin and are highly sensitizing in some people. Formerly, the use of horse antiserum in treatment of infectious diseases led to serious reactions in sensitized persons, but the risk has been substantially reduced in recent years since the advent of human antisera.

Sheep

Little information is available regarding sensitivity to sheep. Major allergens have not been identified. Contact dermatitis, possibly due to lanolin in the wool, can occur (Slavin 1993).

Deer

Some people have been shown to be sensitized to deer proteins. There is evidence of cross sensitivity between deer and horse allergens (Huwyler and Wüthrich 1992). Airborne reindeer epithelial allergens have been detected at 0.1-3.9 µg/m^3 (Reijula and others 1992).

Birds

Exposure to birds can cause rhinitis and asthma symptoms. Birds are also a potential source of hypersensitivity pneumonitis, a lung condition in which a pneumonia-like illness develops after repeated exposure to the antigen. These allergic and hypersensitivity reactions are not mediated by IgE antibody. The symptoms and signs usually occur several hours after exposure and consist of cough, fever, chills, myalgia, and shortness of breath. People with hypersensitivity pneumonitis often have precipitating IgG antibodies to the protein in question. Various bird proteins have been identified as sources of antigens involved in both allergic reactions and hypersensitivity pneumonitis. These proteins are found in pigeon serum and droppings that contain serum.

Reptiles

Human sensitivity to reptiles and amphibians is rare. Cases of occupational asthma caused by frog proteins have been described (Chang-Yeung and Malo 1994), but otherwise the information is sparse.

Fish

Fish proteins are a source of problems for people sensitized through inhalation. In the fish- and crab-processing industry and through the use of fish as a source of animal feed, some people have developed allergic rhinitis and asthma symptoms (Malo and Cartier 1993). Crustaceans and mollusks also pose problems in some laboratory workers. There is evidence that sensitization to airborne allergens from these sources can result in asthma (Malo and Cartier 1993).

Insects

Entomologists are at risk for developing sensitivity to insect proteins. People working in laboratories can be exposed to scales of moths, caterpillars, and other insects that result in sensitization. Beetles, mealworms, cockroaches, and other insects have been described as causing contact urticaria, rhinitis, and possibly asthma symptoms in laboratory workers (Gutman and Bush 1993).

PREVENTIVE MEASURES AND INTERVENTIONS

Prudence suggests that efforts to minimize exposure to animal allergens would result in a reduction in the frequency of sensitization in laboratory-animal workers and a reduction in symptoms in those who have developed sensitivity. But there are few data to support that suggestion. In spite of a number of attempts to reduce or minimize exposure, laboratory-animal allergy remains an important problem. Further research is needed to determine which measures are effective in preventing and controlling symptoms of laboratory-animal allergy.

Screening Programs

Preplacement screening evaluations can be helpful in identifying and alerting persons who might be at risk for developing laboratory-animal allergy or asthma and educating them to take protective measures. The extent of the evaluations depends on the resources of the facility; at a minimum, a simple questionnaire that asks for a personal and family history of allergy (seasonal rhinitis or "hay fever," asthma, eczema, hives) and specifically allergy to laboratory animals (pets, as well as laboratory animals) should be completed. The presence of pre-existing allergic conditions in a person might increase likelihood of development of asthma in an occupational setting where there is exposure to laboratory animals. Because most people will not develop sensitivities beyond pre-existing conditions, this evaluation should not preclude employment. Skin testing or in vitro tests to detect the presence of specific IgE antibodies to animals and other allergens should be available but not required. Positive results can be used to place people with pre-existing sensitivity to laboratory animals in low-risk assignments or used diagnostically to demonstrate the development of sensitization in people who might later become symptomatic. Skin tests are usually applied by the prick or puncture method. Proteins are extracted from the allergen source, usually in a saline buffer, and applied to the skin of the subject, and the skin is pricked with a needle. The demonstration of a wheal and flare response within 10-15 min after application suggests the presence of an IgE-mediated allergic mechanism.

In some instances, a serological immunoassay, such as the radioallergosorbent test (RAST), or an enzyme-linked immunosorbent assay (ELISA) is used to detect the presence of specific IgE antibodies as an alternative to skin tests. In these immunoassays, the person's serum is incubated with the relevant allergenic protein bound to a support material, and the binding of IgE antibodies is detected with a radiolabeled (RAST) or enzyme-linked (ELISA) anti-IgE antibody system. Both skin tests and in vitro assays are reasonably reliable and sensitive in detecting allergy to animal protein, although the quality of materials available commercially for testing is variable.

Clearly, people with pre-existing laboratory animal sensitivity should avoid repetitive exposure. Sensitized people who have had 2 yr or more of experience working with laboratory animals might be at risk for developing airway hyperresponsiveness (asthma) as a result of laboratory animal exposure (Newill and others 1992). In addition to a screening questionnaire for the presence of asthma or asthma symptoms (coughing, wheezing, chest tightness, or shortness of breath), objective measurement of pulmonary function is encouraged. Spirometry, or peak expiratory flow rates, before and after the inhalation of a bronchodilator, are useful in detecting, evaluating the severity of, and monitoring occupation-related asthma.

In people who are chronically exposed to laboratory animals, annual screening should be done to detect those who are developing allergic symptoms (sneezing, nasal congestion, itchy eyes, cough, wheezing, shortness of breath, or hives) so that appropriate intervention measures can be taken to prevent long-term difficulties. Such screening should, at a minimum, consist of a questionnaire regarding allergic or asthma symptoms, and may include skin testing or an in vitro test for specific IgE antibodies to identify sensitization. Periodic monitoring of pulmonary function is recommended if asthma symptoms appear.

Facility Design

Attention to facility design can be helpful in reducing the incidence of laboratory animal allergy. The airborne-allergen load in an animal room depends on the rate of production, which is a function of the numbers of animals present, and the rate of removal, which is a function of ventilation. Achieving a substantial reduction of airborne allergen in a heavily populated area requires extremely high ventilation rates in excess of 100 air changes per hour (Swanson and others 1990). That might be possible with the use of high-efficiency-particulate-air-filtered (HEPA-filtered) laminar-flow units, but such measures can be extremely expensive.

Airborne concentrations of rat allergens also depend on the relative humidity of the environment. An increase in relative humidity from 54% to 77% was shown to reduce airborne rat-allergen concentrations substantially (Edwards and others 1983). This simple maneuver could be of benefit in reducing exposure in some facilities; however, raising humidity to 77% might exceed the optimal range for animals, produce employee discomfort, and induce mold growth.

Cage-emptying where loose bedding is used results in particularly high levels of allergen exposure. Use of ventilated hoods or work stations for cage-emptying and cage-cleaning with filtered, recirculated air can reduce exposure. More detailed discussions of ventilation systems can be found in Hunskaar and Fosse (1993) and Bland and others (1987).

The type of caging will undoubtedly influence exposure to airborne allergens. Filter-top cages have been shown to reduce concentrations of airborne allergens, compared with conventional open-top cages (Gordon and others 1992; Sakaguchi and others 1990). Ventilated cage and rack systems that can reduce exposure are commercially available. Cage and rack systems that exhaust air through a HEPA filter system before returning it into the room substantiatially reduce the concentration of airborne rat allergen, *Rat n 1,* compared with non-HEPA-filtered cage racks (Ziemann and others 1992). However, data to support the routine use of these devices to prevent sensitization or reduce symptoms in workers have not appeared.

Work Practices

Several work practices can reduce the potential development of laboratory animal allergy and perhaps alter its severity. Educational programs and codes of practice can greatly reduce the incidence (37% to 12% over 4 yr) and severity of allergic symptoms (Bothan and others 1987; Olfert 1993). Workers should be made aware of the risks and be instructed in proper measures to control and avoid exposure as much as possible. Those with a history of allergies and particularly those with known sensitivities to animals are at highest risk and so should be especially sought out for education. Sensitized workers who develop asthma should be made aware that they might experience such symptoms not only when exposed to animals but also when they engage in exercise and other physical activities.

The *Guide for the Care and Use of Laboratory Animals* (NRC 1996) recommends that solid-bottom cages with bedding be used for mice and rats. Selection of bedding materials can be beneficial in reducing worker exposure. Use of noncontact absorbent pads, rather than such wood-based contact litter as chips and sawdust, substantially reduced airborne concentrations of rat urinary allergen (Gordon and others 1992).

Job assignment on entry into the laboratory animal work environment should be assessed. People with known risks are best assigned to tasks that minimize exposure. Some tasks—such as simple feeding, weighing, or necropsy—produce low levels of exposure, whereas cage cleaning can lead to high levels of exposure. Selection of job assignment is the first step to minimize exposure of people who have become sensitized or have developed symptoms.

Personal Protective Equipment

The use of protective equipment and clothing can minimize the chance of sensitization. Few data are available to determine which methods are most effective. However, surgical (cloth or paper) disposable masks are probably not effective. The use of gloves, laboratory coats, shoe covers, and other kinds of protective clothing that are worn only in the animal rooms should be encouraged. Frequent hand washing is important and showering after work might be of value.

Once a person develops allergic symptoms, surgical (cloth or paper) disposable masks are usually not effective. Some commercial dust respirators can exclude up to 98% of mouse urinary allergens (Sakaguchi and others 1989b). High-efficiency respirators are most likely to be of value, but they are cumbersome and often are not used appropriately (Hunskaar and Fosse 1993).

At a minimum, for symptomatic workers, the use of a dust-mist respirator certified by the National Institute for Occupational Safety and Health should be required to control symptoms. A filtered airhood device (Airstream Dustmaster® hood, Racal, Middlesex, UK) has been shown to be effective (Price and Longbottom 1988). The use of these devices and protective clothing is most successful in highly motivated workers who have some control over their exposure frequency. Employees using effective respiratory protection (respirators) will need respiratory fit-testing and medical clearance.

EVALUATION OF THE ALLERGIC WORKER

When people develop allergic symptoms (sneezing, nasal congestion, itchy eyes, cough, wheezing, chest tightness, shortness of breath, or hives) related to laboratory animal exposures, consultation with appropriate physicians (allergists, pulmonologists, or occupational medicine specialists) is necessary so that an accurate diagnosis and effective management can be achieved. The American Academy of Allergy, Asthma, and Immunology can provide assistance (AAAAI, 611 East Wells St., Milwaukee, WI 53202. Ph: 414-272-6071; Fax: 276-3349; Web site http://www.AAAAI.org). For personnel in research animal facilities suspected of having allergic problems, the diagnosis of animal sensitivity is based largely on the history of symptoms in conjunction with exposure. The diagnosis is confirmed by the demonstration of specific IgE antibodies to the allergen in question. Pulmonary-function measurements should be done to diagnose or assess asthma severity. Exposure-reduction and -avoidance measures should be undertaken when people become sensitized and develop symptoms resulting from their exposure. Medicines to reduce or prevent allergic or asthma symptoms might be necessary. Many highly sensitized people will continue to have symptoms in spite of exposure reduction and appropriate medications and therefore must avoid animal-allergen exposure completely.

In a few people, immunotherapy against cat and dog allergens has been undertaken with some degree of success (Alvarez-Cuesta and others 1994; Ohman and others 1983). Uncontrolled studies of immunotherapy against allergens of mice, rats, and rabbits have also demonstrated some improvement (Wahn and Siriganian 1980). In general, however, the use of immunotherapy as a means to protect workers from further symptoms has not been fully established.

Further information regarding the evaluation and treatment of workers allergic to laboratory animals can be obtained from professional organizations, such as the American Academy of Allergy, Asthma, and Immunology, the American College of Allergy, Asthma, and Immunology, and the American Thoracic Society.

ANAPHYLAXIS

On rare occasions, an allergic worker might suffer an anaphylactic reaction to an animal bite (Teasdale and others 1993) or from puncture wounds from needles contaminated with laboratory animal protein (Watt and McSharry 1996). These reactions can progress rapidly and become potentially fatal, so physicians might recommend that allergic workers carry a self-administered form of epinephrine (e.g., Epi-Pen® or Ana-Kit®). In appropriate circumstances, it is helpful to instruct co-workers in emergency procedures, such as cardiopulmonary resuscitation.

5

Zoonoses

The transmission of zoonotic disease in the laboratory-animal environment is uncommon, despite the number of animal pathogens that have the capacity to cause disease in humans. That is largely the result of the collaborating interactions and work of two groups. The laboratory-animal industry has had much success in providing high-quality laboratory animals of defined health status for use in research. And research institutions have developed comprehensive and responsive programs of veterinary care that have fostered the investigation of new disease findings and helped to ensure the continuing health of research-animal populations. Quality veterinary care itself, however, is insufficient to prevent the transmission of zoonoses in a research institution. The repeated occurrences of laboratory-acquired Q fever and lymphocytic choriomeningitis and the emergence of newly recognized zoonoses point to a need for investigators to become more involved in their institutions' efforts to prevent occupationally acquired zoonotic disease. The occupational-medicine services might be first to observe the symptoms of zoonotic infection, but it is also important that the institutions' medical professionals become knowledgeable in methods for detecting and managing zoonoses for which workers at the institutions are at risk. All workers share the responsibility for protecting their own health. Personal hygiene affords a critical barrier to the transmission of zoonoses and should be reinforced routinely in an institution's educational efforts and materials, in group and laboratory meetings of involved personnel, and in messages that emphasize appropriate practices for the care and use of research animals.

The following discussion covers most of the zoonotic diseases important to laboratory-animal personnel. The emphasis is on likely occurrence and potential for severity. Some uncommon zoonoses are covered only briefly even though they could have devastating effects if imported into the laboratory environment. In this regard, institutions should investigate situations that are peculiar to proposed research and instructional programs and that might pose special zoonotic hazards—e.g., the use of wild-caught birds or mammals or their fresh carcasses with their associated flora and fauna—before embarking on full-scale programs. That might occasionally necessitate the use of an integrated team from within the institution or of outside specialists or consultants to ensure that the research-animal facilities and personnel expertise are conducive to safety.

The information on zoonotic diseases is organized by agent category. Major sections on viral diseases, rickettsial diseases, bacterial diseases, protozoal diseases, and fungal diseases are included. Material relevant to each zoonotic disease is presented under four headings: reservoir and incidence; mode of transmission; clinical signs, susceptibility, and resistance; and diagnosis and prevention. The discussion on reservoir and incidence addresses the natural infection in the animal host species. The three other headings deal specifically with the potential for and occurrence of occupationally acquired infection of persons involved in the care and use of animals in research.

Various source materials provide detailed information on zoonoses associated with laboratory animals (Fox and Lipman 1991, Fox and others 1984). Readers should find the Centers for Disease Control and Prevention (CDC) *Morbidity and Mortality Weekly Report* indispensable for reviewing contemporary issues pertaining to zoonotic outbreaks.

Although the subject of xenograft transplantation is beyond the scope of this report, vigilance for zoonoses should be an important aspect of all xenograft transplantations. An important consideration should be the potential for exchange of infectious agents between natural and foreign hosts. Xenograft transplantation can inadvertently introduce animal viruses into a new susceptible host. Infection in a new host might not always be apparent. Long-term management of the xenograft recipient is a necessary and prudent practice for maintaining vigilance because new, previously unidentified, pathogens can be anticipated to arise.

VIRAL DISEASES

B-Virus Infection (Cercopithecine herpesvirus 1, CHV1)

Reservoir and Incidence. First described in 1933 (Gay and Holden), B virus produces a life-threatening disease of humans that has resulted in several deaths in the last decade (CDC 1987, 1989a). In macaques, B virus produces a mild clinical disease similar to human herpes simplex. During primary infection, macaques can develop lingual or labial vesicles or ulcers, which generally heal within 1-2 wk. Keratoconjunctivitis or corneal ulcer also might be noted. After acute infection, latency can be established in the ganglia of the sensory nerves serving the region in which virus was introduced. Reactivation of virus from the latent state can result in recurrent viral shedding from peripheral sites and is often associated with physical or psychological stressors, such as ultraviolet irradiation, immunosuppression, disruption of social hierarchy, or other stressful experimental situations (Zwartouw and Boulter 1984). The infection is usually transmitted between macaques via virus-laden secretions through close contact involving primarily the oral, conjunctival, and genital mucous membranes (Weigler 1995).

In a domestic macaque production colony with endemic infection, an age-related increase in the incidence of B-virus infection occurred during adolescence as exposure to the agent continued; the incidence approached 100% in colony-born animals by the end of their first breeding season (Weigler and others 1993). Seroconversion to a B-virus antibody-positive status among wild-caught rhesus monkeys also indicates that eventually 100% of the newly trapped animals acquire the infection. Consequently, B virus should be considered endemic among Asian monkeys of the genus *Macaca* unless the animals have been obtained from specific breeding colonies confirmed to be free of it. Although several species of New World monkeys and Old World monkeys other than members of the genus *Macaca* are known to succumb to fatal B-virus infection, only macaques are known to harbor B virus naturally (Holmes and others 1995).

Mode of Transmission. B virus is transmitted to humans primarily through exposure to contaminated saliva (in bites) and scratches. Transmission related to needlestick injury (Benson and others 1989) and exposure to infected nonhuman-primate tissues (Wells and others 1989) also has occurred. Fomite transmission through an injury obtained in handling contaminated caging was the cause of one identified infection (Palmer 1987). The transmission of B virus by the aerosol route is not thought to be important. Researchers in the field have suggested that asymptomatic human B-virus infection can occur (Benson and others 1989), but it is unknown whether viral reactivation and severe clinical disease can occur later. Human-to-human transmission was recently documented (CDC 1987).

Clinical Signs, Susceptibility, and Resistance. The incubation period between initial exposure and onset of clinical signs ranges from 2 d to about 1 mo, but the time at which symptoms arise after exposure can vary widely. After exposure by bite, scratch, other local trauma, or contamination of vulnerable sites, humans might develop a herpetiform vesicle at the site of inoculation. Early clinical signs and symptoms include myalgia, fever, headache, and fatigue and are followed by progressive neurological disease with numbness, hyperesthesia, paresthesia, diplopia, ataxia, confusion, urinary retention, convulsions, dysphagia, and an ascending flaccid paralysis.

Diagnosis and Prevention. After the outbreak of B-virus infection in monkey handlers in 1987, CDC developed guidelines to prevent it in humans (CDC 1987), which were later revised by Holmes and others (1995). In brief, the recommendations emphasize the need for nonhuman-primate handlers to use protective clothing, including leather gloves and long-sleeved garments for hand and arm protection and face shields or masks and goggles to protect the eyes and mucous membranes from exposure to macaque secretions. Those barrier protections will minimize exposures. The use of latex gloves alone for hand protection should be reserved for the handling of monkeys that are under full chemical restraint. Chemical restraint or specialized restraining devices should be used with nonhuman primates whenever possible to minimize direct contact of personnel with alert monkeys. Despite those handling recommendations and the heightened awareness of the B-virus hazard among personnel, exposure of personnel to monkey bites and scratches remains common, as evidenced by the numerous injuries reported to testing laboratories and CDC (Hilliard 1992). Experimental studies with B virus in animals should be conducted at Animal Biosafety Level 3 (CDC-NIH 1993). Serological methods for the detection of serum antibody are used to diagnose prior exposure to and latent infection with B virus in both humans and animals (Katz and others 1986; Munoz and others 1988). Virus isolation from either the monkey or wound site is also performed, and restriction analysis or the polymerase chain reaction is used later to confirm its presence in any sample that yields a cytopathological result. The CDC recommendations specify that institutions should be prepared to handle patients with a suspect exposure promptly. The wound, if any, should be cleansed thoroughly, and serum samples and cultures should be obtained for serological study and virus isolation from both the patient and the monkey. The initiation of antiviral therapy with acyclovir or ganciclovir might also be warranted if history and symptoms are consistent with B-virus infection. The management of antiviral therapy in B-virus-infected patients is controversial because increasing antibody titer has been demonstrated in a patient after the discontinuation of acyclovir therapy (Holmes and others 1995). Physicians should consult the Viral Exanthems and Herpesvirus Branch, Division of Viral Diseases, Centers for Disease Control and Prevention, Atlanta, GA 30333 (telephone, 404-329-1338) for assistance in case management. Additional information about B-virus diagnostic resources is available through the National Institutes of Health (NIH) B Virus Resource Laboratory, Department of Virology and Immunology, Southwest Foundation for Biomedical Research, P.O. Box 28147, San Antonio, TX 78228 (telephone, 210-674-1410).

Ebola-Virus Infection

Reservoir and Incidence. Ebola hemorrhagic fever is a rare disease caused by a filovirus that is structurally identical with, but antigenically distinct from, Marburg-disease virus. Cases of disease related to this agent have been restricted to the continent of Africa. Sudan and Zaire strains of the virus have been shown experimentally to produce lethal infection in nonhuman primates in about 8 d, but monkeys have not been shown to be the natural reservoir (Dalgard and others 1992; Johnson 1990a); the natural reservoir for Ebola virus has not yet been identified.

The identification and isolation of an Ebola-like filovirus, Ebola-Reston, from macaques imported into the United States from the Philippines during 1989, the first appearance of an Ebola viral strain that did not originate in the continent of Africa, prompted the implementation of revised nonhuman-primate importation and handling guidelines (CDC 1989b, 1990). Although Ebola-Reston was less virulent than Ebola-Zaire or Ebola-Sudan in nonhuman primates, it also produced a hemorrhagic disease that involved multiple organ systems and produced death in 8-14 d in infected macaques. The natural reservoir of the Ebola-Reston strain has not been determined. However, a new strain of Ebola virus has been isolated from naturally infected chimpanzees from a wild troop that had experienced outbreaks of disease characterized by a hemorrhagic syndrome. Further study of this troop might begin to resolve questions about the natural reservoirs of the Ebola virus (Le Guenno and others 1995).

Mode of Transmission. Transmission of Ebola-virus infection during epidemics among humans generally has involved close contact, and the low secondary-attack rate suggests that transmission is not efficient (Murphy and others 1990). Sexual contact and nosocomial transmission through exposure to contaminated syringes and needles, infected tissues, blood, and other bodily fluids are important means of viral transmission. Aerosol transmission has not been a feature of the African Ebola-virus outbreaks to date, but it cannot be discounted completely. During the outbreak of Ebola-Reston disease in the nonhuman-primate colonies in the United States, its spread within rooms between animals without direct contact supported the possibility of droplet or aerosol transmission.

Clinical Signs, Susceptibility, and Resistance. In humans, the Zaire and Sudan strains produce a disease characterized by multifocal organ necrosis, coagulopathy, extensive visceral effusions, hemorrhagic shock, and death. Human infections with the Reston strain during the outbreak in nonhuman primates were subclinical but resulted in seroconversion.

Diagnosis and Prevention. A wide variety of techniques can be used to detect Ebola virus or the viral antigen. The infection is diagnosed serologically on the basis of antibody titer in indirect immunofluorescence assay, radioimmunoassay, and enzyme-linked immunosorbent assay.

The CDC-mandated procedures for importation of nonhuman primates limit the occurrence of this disease to facilities involved in importation (CDC 1990). Personnel in those facilities should become familiar with the specialized equipment and procedures used to minimize Ebola-virus transmission in the event of an outbreak. Neither vaccines nor therapeutic pharmaceuticals are available for the prevention or treatment of Ebola-virus infection. The Subcommittee on Arbovirus Laboratory Safety (SALS) of the American Committee on Arthropod-Borne Viruses recommends that work with Ebola virus be conducted at the equivalent of Biosafety Level 4 (CDC-NIH 1993).

Marburg-Virus Disease

Reservoir and Incidence. Marburg-virus disease has been recognized on only four occasions. The index cases involved 31 persons in three European laboratories who were handling tissues from African green monkeys; seven of the 31 died (Martini and Siegert 1971). There was no secondary spread of the disease among the monkeys in the facility, and no infections occurred among the animal-care staff (Martini 1973). Although African green monkeys, other nonhuman primates, and other animals are susceptible and succumb to fatal infection, the natural reservoir for the virus has not been determined (Benenson 1995a; Simpson and others 1968).

Mode of Transmission. The transmission of Marburg virus from animals to humans has involved direct contact with infected tissues. Aerosol transmission has been suggested as a means of transmission among monkeys (Hunt and others 1978). Person-to-person transmission occurs by direct contact with contaminated blood, secretions, organs, or semen.

Clinical Signs, Susceptibility, and Resistance. Marburg virus produces a serious disease, and apparently everyone is susceptible to it. After an incubation period of 4-16 d, humans develop fever, myalgia, headache, and conjunctival suffusion. Nausea, vomiting, and severe diarrhea appear within 2-3 d with thrombocytopenia and leukopenia. Other organ involvement can include pancreatitis, orchitis, hepatocellular necrosis, and a maculopapular rash. Abnormalities in the coagulation pattern indicative of disseminated intravascular coagulation occur and might be the proximate cause of death in one-fourth of the cases.

Diagnosis and Prevention. The diagnosis of Marburg-virus infection depends primarily on isolation of the virus from blood or tissue specimens. Immunofluorescent staining has demonstrated viral antigen in tissue samples with high concentrations of infectious materials. An immunofluorescence assay also has been developed to detect serum antibodies in recovering patients (Fox and Lipman 1991; Jahrling 1989).

SALS recommends that work with Marburg virus be conducted at the equivalent of Biosafety Level 4 (CDC-NIH 1993).

Hantavirus Infection (Hemorrhagic Fever with Renal Syndrome and Nephropathia Endemica)

Reservoir and Incidence. Hantavirus is one of several genera in the family Bunyaviridae that can cause severe hemorrhagic disease. The hantaviruses are widely distributed in nature among wild-rodent reservoirs, and the severity of the disease produced depends on the virulence of the strain involved (Gajdusek 1982; LeDuc 1987). Strains producing hemorrhagic fever with renal syndrome are prevalent in southeastern Asia and Japan and focally throughout Eurasia. Strains producing a less-severe form of the disease known as nephropathia endemica occur throughout Scandinavia, Europe, and western portions of the former Soviet Union. Outbreaks of hantavirus infection characterized by a severe pulmonary syndrome resulting in numerous deaths were recently recognized in the southwestern United States (CDC 1993a,b; CDC 1995, CDC 1996).

Rodents in numerous genera (*Apodemus*, *Clethrionomys*, *Mus*, *Rattus*, *Pitimys*, and *Microtus*) have been implicated in foreign outbreaks of the disease. In the United States, serological surveys have detected evidence of hantavirus infection in urban and rural areas involving the following rodents: *Rattus norvegicus*, *Peromyscus* spp., *Microtus californicus*, *Tamias* spp., and *Neotoma* spp. (CDC 1993a,b; Tsai and others 1985). Numerous cases of hantavirus infection have occurred in laboratory animal facility people from exposure to infected rats (*Rattus*), including outbreaks in Korea, Japan, Belgium, France, and England (LeDuc 1987). There is also epidemiologic evidence that cats can become infected through rodent contact and potentially serve as a reservoir (Xu and others 1987).

Mode of Transmission. The transmission of hantavirus infection is through the inhalation of infectious aerosols, and extremely brief exposure times (5 min) have resulted in human infection. Rodents shed the virus in their respiratory secretions, saliva, urine, and feces for many months (Tsai 1987). Transmission of the infection also can occur by animal bite or when dried materials contaminated with rodent excreta are disturbed, allowing wound contamination, conjunctival exposure, or ingestion to occur (CDC 1993a,b). The recent cases that have occurred in the laboratory-animal environment have involved infected laboratory rats. In such an environment, the possibility of transmitting the infection from animal to animal by the transplantation of cells or tissues also should be considered (Kawamata and others 1987). Person-to-person transmission apparently is not a feature of hantavirus infection.

Clinical Signs, Susceptibility, and Resistance. The clinical signs are related to the strain of hantavirus involved. The form of the disease known as nephropathia endemica is characterized by fever, back pain, and a nephritis that causes only moderate renal dysfunction, from which the patient recovers; in the recent cases in the United States, patients had fever, myalgia, headache, and cough followed by rapid respiratory failure (CDC 1993a,b). The form of the disease that has been noted after laboratory-animal exposure fits the classical pattern of hemorrhagic fever with renal syndrome; the infection is characterized by fever, headache, myalgia, and petechiae and other hemorrhagic manifestations, including anemia, gastrointestinal bleeding, oliguria, hematuria, severe electrolyte abnormalities, and shock (Lee and Johnson 1982).

Diagnosis and Prevention. Human hantavirus infections associated with the care and use of laboratory animals should be prevented through the isolation or elimination of infected rodents and rodent tissues before they can be introduced into resident laboratory-animal populations. Serodiagnostic tests are available for both animals and humans. Additional information about serological testing is available through the Special Pathogens Branch, Division of Viral and Rickettsial Diseases, National Center for Infectious Diseases, CDC. Rodent tumors and cell lines can be tested for hantavirus contamination with a modified rat-antibody production test. People suspected of having the infection might benefit from intravenous ribavirin therapy initiated early in the course of the disease (Morrison and Rathbun 1995). Hemodynamic maintenance and respiratory support are critical for these people after infection.

Animal Biosafety Level 2 is recommended for working with experimentally infected rodent species known not to excrete the virus. All work involving inoculation of the virus into *P. maniculatus* or other permissive species should be conducted at Animal Biosafety Level 4 (CDC 1994b).

Lymphocytic Choriomeningitis Virus Infection

Reservoir and Incidence. Lymphocytic choriomeningitis (LCM) virus is a member of the family Arenaviridae, which consists of single-stranded-RNA viruses with a predilection for rodent reservoirs. Several important zoonoses are associated with this family, including Lassa fever and Argentine and Bolivian hemorrhagic fevers, but only LCM is important as a natural infection of laboratory animals. Human infection with LCM associated with laboratory-animal and pet contact has been recorded on numerous occasions (Fox and others 1984; Jahrling and Peters 1992). LCM is widely distributed among wild mice throughout most of the world and presents a zoonotic hazard. Many laboratory-animal species are infected naturally, including mice, hamsters, guinea pigs, nonhuman primates, swine, and dogs; but the mouse has remained the species of primary concern in the consideration of this disease, as it was in a recent outbreak of LCM in humans (Dykewicz and others 1992). Athymic, severe-combined-immunodeficiency (SCID), and other immunodeficient mice can pose a special risk of harboring silent, chronic infections and present a hazard to personnel (CDC-NIH 1993; Dykewicz and others 1992).

Mode of Transmission. The LCM virus produces a pantropic infection under some circumstances and can be present in blood, cerebrospinal fluid, urine, nasopharyngeal secretions, feces, and tissues of infected natural hosts and possibly humans. Bedding material and other fomites contaminated by LCM-infected animals are potential sources of infection, as are infected ectoparasites. In endemically infected mouse and hamster colonies, the virus is transmitted in utero or early in the neonatal period and produces a tolerant infection characterized by chronic viremia and viruria without marked clinical disease; spread of LCM among animals via contaminated tumors and cell lines also should be recognized (Bhatt and others 1986; Nicklas and others 1993). Infection in humans can be by parenteral inoculation, inhalation, and contamination of mucous membranes or broken skin with infectious tissues or fluids from infected animals. Aerosol transmission is well documented. The virus can pose a special risk during pregnancy: that of infection of the fetus.

Clinical Signs, Susceptibility, and Resistance. Humans develop an influenza-like illness characterized by fever, myalgia, headache, and malaise after an incubation period of 1-3 wk. In severe cases of the disease, patients might develop a maculopapular rash, lymphadenopathy, meningoencephalitis, and, rarely, orchitis, arthritis, and epicarditis (Johnson 1990b). Central nervous system involvement has resulted in several deaths (Benenson 1995b).

Diagnosis and Prevention. Virus isolation from blood or spinal fluid in conjunction with immunofluorescence assay of inoculated cell cultures is the main method of diagnosing acute disease. Antibody is detectable with such an assay about 2 wk after the onset of illness. Prevention of this disease in the laboratory is achieved through the periodic serological surveillance of new animals that have inadequate disease profiles and of resident animal colonies at risk and through screening for the presence of LCM in all tumors and cell lines intended for animal passage. Intravenous ribavirin therapy substantially reduces mortality in patients infected with Lassa fever virus and also might be useful for LCM virus (Andrei and De Clercq 1993). Additional information about therapy and serological testing for LCM is available through the Special Pathogens Branch, Division of Viral and Rickettsial Diseases, National Center for Infectious Diseases, CDC.

Animal Biosafety Level 2 is recommended for studies in adult mice with mouse brain-passage strains. Animal Biosafety Level 3 should be used for work with infected hamsters (CDC-NIH 1993).

Poxvirus Diseases of Nonhuman Primates
(Monkeypox and Benign Epidermal Monkeypox)

Reservoir and Incidence. Monkeypox is an orthopoxvirus closely related to smallpox and produces a clinical disease similar to smallpox. Sporadic cases of the human disease are noted in Africa. Recently, squirrels of the genera *Funisciurus* and *Heliosciurus* have been identified as hosts and important reservoirs of the virus (Benenson 1995b). Natural outbreaks of monkeypox also have been recorded in nonhuman primates in the wild and laboratory settings (Fox and others 1984).

Benign epidermal monkeypox, or tanapox, is a poxvirus that affects monkeys of the genus *Presbytis* in Africa and captive macaques in the United States.

Mode of Transmission. The transmission of monkeypox from laboratory nonhuman-primate populations to humans has not been recorded. Human-to-human transmission of the agent has occurred, presumably through close contact with active lesions, recently contaminated fomites, or respiratory secretions. The possibility of zoonotic spread should be considered.

Benign epidermal monkeypox has been transmitted from monkeys to humans in the laboratory-animal environment (McNulty 1968). Direct contact with infected animals or contaminated fomites is necessary for disease transmission.

Clinical Signs, Susceptibility, and Resistance. Monkeypox is of interest and importance primarily because it produces a disease similar to smallpox characterized by fever, malaise, headache, severe backache, prostration, and occasional abdominal pain. Lymphadenopathy and a maculopustular rash develop later. Some patients develop a severe fulminating disease and die.

Benign epidermal monkeypox is characterized by the development of circumscribed, oval to circular, raised red lesions usually on the eyelids, face, body, or genitalia. The lesions regress spontaneously in 4-6 wk.

Diagnosis and Prevention. The diagnosis of poxvirus infections can be established on the basis of the characteristic structure of viral particles as seen with the electron microscope. Virus isolation on chick chorioallantoic membrane and characterization with specific biological tests are needed to differentiate among the various orthopoxviruses. Vaccinia vaccination is protective against monkeypox in humans and monkeys (Benenson 1995b).

Orf Disease (Contagious Ecthyma and Contagious Pustular Dermatitis)

Reservoir and Incidence. Orf disease is a poxvirus infection that is endemic in many sheep flocks and goat herds throughout the United States and worldwide. The disease affects all age groups, although young animals are most often and most severely affected. Orf produces proliferative, pustular encrustations on the lips, nostrils, mucous membranes of the oral cavity, and urogenital orifices of infected animals (Fox and others 1984).

Mode of Transmission. Orf, a double-stranded-DNA virus, is transmitted to humans by direct contact with virus-laden lesion exudates. External lesions are not always apparent, so recognition can be difficult. Transmission of the agent by fomites or contaminated animals is possible because of its environmental persistence. Rare cases of person-to-person transmission have been recorded (Benenson 1995b).

Clinical Signs, Susceptibility, and Resistance. The disease in humans is usually characterized by the development of a solitary lesion on the hand, arm, or face. The lesion is initially maculopapular or pustular and progresses to a weeping proliferative nodule with central umbilication. Such lesions are sometimes mistaken for abscesses but should not be lanced. Occasionally, several nodules are present, each measuring up to 3 cm in diameter, persisting for 3-6 wk and regressing spontaneously. Regional adenitis is uncommon, and progression to generalized disease is considered rare (Erickson and others 1975).

Diagnosis and Prevention. The characteristic appearance of the lesion and a history of recent contact with sheep or goats are diagnostic of this condition in humans. Vaccination of susceptible sheep and goats is effective in preventing the disease. Personnel who handle sheep and goats should be cautioned to wear protective clothing and gloves and to practice good personal hygiene.

Measles (Rubeola)

Reservoir and Incidence. Humans are the reservoir for measles. Nonhuman primates become infected through contact with human populations with endemic measles (Fox and others 1984). Both Old World and New World nonhuman primates are susceptible to infection (Fox and others 1984). The disease spreads rapidly through infected nonhuman-primate colonies; wild-caught nonhuman-primate populations often attain a 100% seroconversion rate within several weeks of capture. However, with the current emphasis on and success of domestic nonhuman-primate production, institutions could develop large populations of susceptible nonhuman primates.

Mode of Transmission. Measles, a highly communicable disease, is transmitted via infectious aerosols, contact with nasal or throat secretions, or contact with fomites freshly contaminated with infectious secretions.

Clinical Signs, Susceptibility, and Resistance. The clinical signs of measles are similar in nonhuman primates and humans. In humans, fever develops after an incubation period of about 10 d and is followed by conjunctivitis, coryza, cough, and Koplik's spots inside the mouth. Later, a characteristic exanthematous rash develops, beginning on the face, becoming generalized over the body, and ending sometimes in flaky desquamation. Complications of viral replication or secondary bacterial infection can result in pneumonia, otitis media, diarrhea, or, rarely, encephalitis (Benenson 1995b).

Diagnosis and Prevention. Characteristic clinical signs generally make diagnostic methods unnecessary, but serology, immunofluorescent-antibody screening for virus in clinical specimens, or viral isolation can be used. Vaccination of all nonhuman-primate handlers against measles should be ensured, and vaccination of nonhuman-primate populations also should be considered.

Newcastle Disease

Reservoir and Incidence. Newcastle disease is caused by a paramyxovirus. It is seen among wild, pet, and domestic birds, and wild birds transmit the infection to domestic-bird populations (Bryant 1984). The zoonotic potential of the agent in the laboratory environment has been realized on numerous occasions (Barkley and Richardson 1984).

Mode of Transmission. Aerosol transmission is the important means of spread, but contaminated food, water, and equipment also transmit infection within bird populations.

Clinical Signs, Susceptibility, and Resistance. The severity of the disease in birds depends on the pathogenicity of the infecting strain. Highly pathogenic strains have been largely excluded from flocks within the United States. Moderately pathogenic strains produce anorexia and respiratory disease in adult birds and neurological signs in young birds. The disease in humans is characterized by follicular conjunctivitis, mild fever, and respiratory involvement ranging from cough to bronchiolitis and pneumonia.

Diagnosis and Prevention. In the laboratory environment, the disease can be prevented by immunizing susceptible birds against it or obtaining birds from flocks known to be free of the agent. Good personal-hygiene practices also should be in place.

Hepatitis A

Reservoir and Incidence. Humans are the primary reservoir for hepatitis A virus (HAV), and nonhuman-primate infections result from contact with infected humans. However, more than 200 cases of HAV infection in humans have been associated with nonhuman primates (Barkley and Richardson 1984), and many nonhuman-primate species are susceptible, including chimpanzees and other great apes, marmosets, owl monkeys, cynomolgus monkeys, and patas monkeys (Fox and Lipman 1991; Hollinger and Glombicki 1990). A recent outbreak of HAV infection in young, domestically reared rhesus monkeys has renewed the concern for potential zoonotic transmission (Lankas and Jensen 1987).

Mode of Transmission. HAV is transmitted by the fecal-oral route, and some outbreaks can be related to contaminated food and water.

Clinical Signs, Susceptibility, and Resistance. The disease in nonhuman primates is much less severe than the disease in humans and is often subclinical. Some species of nonhuman primates develop malaise, vomiting, jaundice, and increased serum concentrations of hepatic enzymes.

The disease in humans varies from a mild illness lasting 1-2 wk to a severely debilitating illness lasting several months. After an incubation period of about a month, patients experience an abrupt onset of fever, malaise, anorexia, nausea, and abdominal discomfort, followed within a few days by jaundice. Children often have mild disease without jaundice, whereas HAV infections in older patients can be fulminant and protracted with prolonged convalescence.

Diagnosis and Prevention. Enzyme immunoassay and radioimmunoassay are available for the demonstration of immunoglobulin M-specific anti-HAV in the serum or plasma. Alternatively, fecal samples can be tested for virus particles or viral antigen.

An approved vaccine is now available for the control of HAV infection in humans. Passive immunization with immune serum globulin has also been used at intervals of 4-6 mo for personnel in contact with recently imported chimpanzees (Fox and Lipman 1991). The use of protective clothing, good personal hygiene, and appropriate practices of sanitation of equipment and facilities also will minimize the potential for zoonotic transmission.

Hepatitis B, C, D, and E

Humans are considered the natural host for hepatitis B, C, D, and E viruses (Benenson 1995b). Various nonhuman primates, particularly chimpanzees, can be infected experimentally, but only one case of natural infection has been reported (Kornegay and others 1985). Viral hepatitis B has been suggested in recently imported cynomolgus monkeys by the demonstration of hepatitis B surface antigen in hepatic cells (Kornegay and others 1985), but it was not associated with zoonotic disease transmission; these animals developed mild clinical disease characterized by anorexia, increased hepatic enzyme concentrations, and hyperbilirubinemia. Although natural infections of nonhuman primates with hepatitis B, C, D, and E viruses are extremely rare, personnel should adhere to appropriate precautions when handling nonhuman primates.

Animal Biosafety Level 2 practices, containment equipment, and facilities are recommended for activities using naturally or experimentally infected chimpanzees or other nonhuman primates. Licensed recombinant vaccines against hepatitis B are available and highly recommended for personnel involved in studies with hepatitis B virus (CDC-NIH 1993).

Simian Immunodeficiency Virus (SIV) Infection

Reservoir and Incidence. Simian immunodeficiency virus (SIV) is a lentivirus that produces in rhesus monkeys and other susceptible macaque species a clinical syndrome that has many important parallels to AIDS. Although the seroprevalence of SIV in Asian macaques is low and most SIV infections in these species are related to their use as animal models of AIDS, the seroprevalence among wild-caught African green monkeys (*Cercopithecus aethiops*), which apparently does not manifest clinical signs, is about 30% or higher (Lairmore and others 1989).

Mode of Transmission. Transmission of SIV between monkeys is believed to require direct inoculation of open wounds or mucous membranes with infectious secretions. Aerosol transmission between monkeys has not been demonstrated where uninfected macaques have been housed in separate cages near SIV-infected monkeys (Lairmore and others 1989). The blood, secretions, and tissues of SIV-infected monkeys should be presumed to be infectious for persons potentially exposed to these materials. Two human cases of seroconversion associated with known exposure have been recognized (CDC 1992a; Khabbaz and others 1992), and a blind serological survey of other personnel working with SIV has identified perhaps an additional three seropositive persons. The possible inclusion of the aforementioned cases of known SIV exposure and the cross reactivity of SIV and HIV-2 in the assay used confounded the interpretation of the results of this survey (CDC 1992b). The person involved in the first case had a skin puncture caused accidentally by a needle contaminated by the blood of an infected macaque. In the second case, a laboratory worker who had hand and forearm dermatitis handled SIV-infected blood specimens without wearing gloves. The pattern of seroreactivity suggested the possibility of infection in the second case, and attempts to isolate SIV from this person were successful (Khabbaz and others 1992).

Clinical Signs, Susceptibility, and Resistance. Clinical signs have not been recorded in cases of human SIV exposure.

Diagnosis and Prevention. Serological techniques and virus isolation are available for the diagnosis of SIV exposure and infection. Personnel should be enrolled in a medical-surveillance program and maintain work practices consistent with the handling of bloodborne pathogens (CDC 1988). Animal Biosafety Level 2 practices, containment equipment, and facilities are recommended for activities using naturally or experimentally infected nonhuman primates or other animals.

Rabies

Reservoir and Incidence. Rabies occurs worldwide except for a few countries that have excluded the disease through animal-importation and animal-control programs and the aid of geographic barriers (Fox and others 1984). Rabies virus infects all mammals, but the main reservoirs are wild and domestic canines, cats, skunks, raccoons, bats, and other biting animals. The disease historically has not posed a problem in the laboratory-animal setting. However, the incidence of rabies in wildlife in the United States has been rising in recent years, and the possibility of rabies transmission to dogs or cats with uncertain vaccination histories and originating in an uncontrolled environment must be considered. In addition, rabies-susceptible wildlife introduced into the laboratory for special research investigations have the potential to harbor infection.

Mode of Transmission. Rabies virus is most commonly transmitted by the bite of a rabid animal or the introduction of virus-laden saliva into a fresh skin wound or an intact mucous membrane. Airborne transmission probably can occur in caves where rabid bats roost, but this mode of transmission is extremely unlikely in the laboratory (Benenson 1995b). The virus also has been transmitted through corneal transplants from persons with undiagnosed central nervous system disease. Personnel who handle tissue specimens or other materials potentially laden with rabies virus during necropsy or other procedures should be regarded as at risk for infection.

Clinical Signs, Susceptibility, and Resistance. Rabies produces an almost invariably fatal acute viral encephalomyelitis. Patients experience a period of apprehension and develop headache, malaise, fever, and indefinite sensory changes referred to the site of a prior animal-bite wound. Further progression of the disease is marked by paresis or paralysis, inability to swallow and the related hydrophobia, delirium, convulsions, and coma. Death is often due to respiratory paralysis.

Diagnosis and Prevention. Rabies usually is diagnosed with specific immunofluorescent antibody staining of brain tissue, corneal smears, mucosal scrapings, or frozen skin-biopsy specimens. Virus isolation also can be used to confirm the diagnosis. The most important factor in preventing human rabies, apart from the immediate and thorough cleaning of bite and scratch wounds, is control of the disease in the domestic-animal population. Stringent vaccination measures and enforced animal-control measures help to reduce the population at risk. Whenever possible, animals brought into the laboratory should have histories that preclude their exposure to rabies or ensure their having been vaccinated for this disease. Pre-exposure immunization should be available to personnel in high-risk categories, such as veterinarians, people who are working with or involved in the care of infected or inadequately characterized animals, and wildlife-conservation personnel who work in rabies-endemic areas. Animal Biosafety Level 2 practices, containment equipment, and facilities are recommended for activities using naturally or experimentally infected animals (CDC-NIH 1993).

Influenza

Reservoir and Incidence. Humans are considered the reservoir for human-influenza viruses. Influenza-virus infections with different antigenic strains occur naturally in many animals, including avian species, swine, horses, mink, and seals (Benenson 1995b). Animal reservoirs are thought to contribute to the emergence of new human strains of influenza viruses, perhaps by reassortment of animal strains with human strains. In the laboratory, ferrets are highly susceptible to human influenza and often are used as experimental models of influenza (Fox and Lipman 1991).

Mode of Transmission. Transmission is by the airborne route and by direct contact. The transmission of animal-influenza strains from animals to humans is rare (CDC-NIH 1993). However, ferrets housed in the laboratory will develop epizootic infection concomitant with human outbreaks of the disease. Ferret-to-human transmission of the virus also has been documented (Marini and others 1989).

Clinical Signs, Susceptibility, and Resistance. Influenza is an acute disease of the respiratory tract characterized by fever, headache, myalgia, prostration, coryza, sore throat, and cough. Viral pneumonia and gastrointestinal involvement manifested by nausea, vomiting, and diarrhea also can develop.

Diagnosis and Prevention. Personnel should wear appropriate protective clothing and practice good personal hygiene if contact with ferrets suspected of having influenza is unavoidable.

Arboviral Infection

Reservoir and Incidence. The arboviruses (arthropod-borne viruses) are taxonomically diverse, each involving its own web of mammalian or avian hosts (or both) and specific arthropod vectors (Benenson 1995b; Tsai 1991). The presence of arboviral infection among laboratory animals generally would be restricted to situations where these agents are the focus of experimental study, wild-caught animals are brought into the laboratory for study, or nontraditional laboratory animals are housed outdoors, permitting the perpetuation of the natural cycle of arboviral infection.

Mode of Transmission. Natural cycles of infection involve transmission from mosquitoes, ticks, midges, or sandflies (Benenson 1995b; Tsai 1991). In the laboratory setting, transmission can occur via parenteral inoculation, aerosol exposure, contamination of unprotected broken skin, and possibly animal bites (CDC-NIH 1993).

Clinical Signs, Susceptibility, and Resistance. The clinical manifestations of arboviral infections are diverse, including fever, hemorrhagic fever, rash, arthralgia, arthritis, meningitis, and encephalitis (Benenson 1995b).

Diagnosis and Prevention. Personnel involved in research-animal studies of arboviral infections should observe strictly the biosafety-level practices deemed appropriate for the particular arboviral agent (CDC-NIH 1993, SALS 1980). Institutions sponsoring research programs involving wild-caught animals should ensure that veterinary and occupational-health personnel have performed an adequate review of the scientific literature to establish a potential-disease profile for the animal species under study and have implemented corresponding measures for personnel protection.

RICKETTSIAL DISEASES

Q Fever

Reservoir and Incidence. Q fever is caused by the rickettsial agent *Coxiella burnetii*. *C. burnetii* has a worldwide distribution perpetuated in two intersecting cycles of infection—in domestic animals and in wildlife animals and their associated ticks. Infection is widespread within the domestic-animal cycle, which includes sheep, goats, and cattle. Cats, dogs, and domestic fowl also can be infected (Fox and others 1984). The prevalence of the infection among sheep is high throughout the United States, and sheep have been the primary species associated with outbreaks of the disease in laboratory-animal facilities (Bernard and others 1982). However, an outbreak of Q fever with one death in a human cohort exposed to a parturient cat and her litter and cases of the disease associated with exposure to rabbits indicate that other species should not be overlooked as possible sources of the infection in the laboratory environment (Langley and others 1988; Marrie and others 1990).

Mode of Transmission. Humans usually acquire this infection via inhalation of infectious aerosols, although transmission by ingestion has been recorded (Benenson 1995b). The organism is shed in urine, feces, milk, and especially birth products of domestic ungulates, which generally are asymptomatic. The placenta of an infected ewe can contain up to 10^9 organisms per gram of tissue, and milk can contain 10^5 organisms per gram (CDC-NIH 1993). The organism is resistant to desiccation and persists in the environment for long periods, contributing to the widespread dissemination of infectious aerosols. The risk of infection is high because the infectious dose by inhalation is less than 10 microorganisms (CDC-NIH 1993, Wedum and others 1972). The importance of those factors was evident in outbreaks of the disease associated with the use of pregnant sheep in research facilities in the United States when personnel became infected along the routes of sheep transport and in the vicinity of sheep surgery from contact with soiled linens (Bernard and others 1982).

Clinical Signs, Susceptibility, and Resistance. The disease in humans varies widely in duration and severity, and asymptomatic infection is possible. The disease often has a sudden onset with fever, chills, retrobulbar headache, weakness, malaise, and profuse sweating. In some cases, pneumonitis occurs with a nonproductive cough, chest pain, and few other signs. Acute pericarditis and acute or chronic granulomatous hepatitis also have been reported. Endocarditis can occur on native or prosthetic cardiac valves and often extends over a period of months or years and results in relapsing systemic infection. Most cases of Q fever resolve within 2 wk (Benenson 1995b). Persons with valvular heart disease should not work with *C. burnetii* (CDC-NIH 1993).

Diagnosis and Prevention. Serological methods available for the detection of a rise in specific antibody between acute and convalescent samples include microagglutination, immunofluorescent, complement fixation (CF), and ELISA tests. The organism can be isolated from blood or other tissues, but doing so poses a hazard for laboratory personnel.

Recommendations for the control of Q fever in a research facility are available and should be applied rigorously in surgical, laboratory, and housing areas used for sheep (Bernard and others 1982). In brief, the recommendations emphasize the need for the separation of sheep-research activities from other areas. Physical barriers or air-handling systems, the appropriate use and disposal of protective clothing, and the use of disinfectants in the sanitation and waste-management programs minimize the risk of exposure. Whenever possible, male or nonpregnant female sheep should be used in research programs. However, many research studies require the use of pregnant sheep. Neither antimicrobial therapy nor serological testing in combination with the culling of infected animals has led to the reliable development of disease-free flocks for use in biomedical-research programs (Fox and Lipman 1991). Serological monitoring of sheep for evidence of *C. burnetii* infection also is unrewarding because serological status is not a useful indicator of organism shedding. Since infected guinea pigs and other rodents may shed the organism in urine and feces, the CDC and NIH recommend maintaining experimentally infected rodents under Animal Biosafety Level 3 (CDC-NIH 1993).

An investigational new Phase 1 Q-fever vaccine is available from the Special Immunizations Program, US Army Medical Research Institute for Infectious Disease (USAMRIID), Fort Detrick, Maryland 21701. The use of this vaccine should be limited to personnel at high risk of exposure who have no demonstrated sensitivity to Q-fever antigen.

Cat-Scratch Fever

Reservoir and Incidence. Bartonella henselae, a newly described rickettsial organism, has been directly associated with cat-scratch fever and bacillary angiomatosis, an unrelated condition that develops usually in people infected with the human immunodeficiency virus (Koehler and others 1994). This gram-negative, pleomorphic organism has a predilection for intracellular growth and has been demonstrated to produce chronic, asymptomatic bacteremia, especially in younger cats, for at least 2.5 mo and possibly as many as 17 mo. The organism has been isolated on fleas that fed on infected cats, and fleas have been shown to be capable of transmitting the organism between cats. This finding suggests that fleas could serve as a vector in zoonotic transmission (Chomel and others 1996). Results of a recent prevalence survey indicated that about 40% of pet and pound cats examined had blood cultures positive for the organism and six of 13 households with cats had at least one positive cat (Koehler and others 1994). Although cat-scratch fever usually has been associated with the scratch or bite of a young cat, other animals have been implicated, including dogs, monkeys, and porcupines (Goldstein 1990b). The incidence of the disease in humans is unknown; an estimate of 2.5 cases per 100,000 population per year has been proposed (Groves and others 1993).

Mode of Transmission. Of patients with the disease, 75% report having been bitten or scratched by a cat, and over 90% report a history of exposure to a cat. Most cases of the disease appear between September and February, and the incidence peaks in December (Fox and others 1984).

Clinical Signs, Susceptibility, and Resistance. The disease begins with inoculation of the organism into the skin of an extremity, usually a hand or forearm. A small erythematous papule appears at the site of inoculation several days later and is followed by vesicle and scab formation. The lesion resolves within a few days to a week. Several weeks later, regional lymphadenopathy appears, often in a solitary lymph node, and it can persist for months. Suppuration of the lymph node sometimes occurs. Fever, malaise, anorexia, headache, and splenomegaly can also be present. Other, less-frequent complications of the disease include periocular lymphadenopathy with palpebral conjunctivitis, central nervous system involvement, osteolytic lesions, granulomatous hepatitis, and pneumonia. Cat-scratch fever can progress to a severe systemic or recurrent infection that is life-threatening in immunocompromised hosts. Such severe cases are reminiscent of bacillary angiomatosis, a condition of HIV-infected patients.

Diagnosis and Prevention. Isolation of the causative organism from the blood, a cutaneous lesion, or biopsy material is required for a definitive diagnosis of cat-scratch fever. Clinical signs, a history of cat contact, failure to isolate other bacteria from affected tissues, and histopathological examination of lymph-node biopsies are used for diagnosis by most physicians (Groves and others 1993). Many patients can be found to be serologically positive for *R. henselae* with the indirect fluorescent-antibody test.

The use of proper cat-handling techniques and protective clothing should minimize the likelihood of personnel exposure to the organism of cat-scratch fever. Clinical trials have indicated that antibiotic treatment can be used to eliminate the carrier state in cats (Koehler and others 1994), but this approach to disease prevention might be impeded by the current difficulty in detecting the carrier state. Flea-control measures should also be implemented.

Other Rickettsial Diseases

Reservoir and Incidence. Dogs, rodents, and their ticks and fleas are the reservoirs for *Rickettsia rickettsia. R. akari, R. prowazekii,* and *R. typhi* are found in wild rodents and their associated fleas and mites (Fox and others 1984). *Ehrlichia canis* produces natural infection only in dogs; human infections result from the bites of infected ticks. These rickettsial infections are considered rare in the United States.

Mode of Transmission. Zoonotic transmission of these diseases in the laboratory has involved aerosols, accidental parenteral inoculation, and bites by natural ectoparasitic vectors (CDC-NIH 1993).

Clinical Signs, Susceptibility, and Resistance. These rickettsial diseases are characterized by fever, headache with encephalitis, myalgia, and a rash of varied distribution according to the species involved (Saah 1990). A rash does not develop in *E. canis* infections. Eschar development at the site of a vector bite is seen in *R. rickettsia* and *R. akari* infections.

Diagnosis and Prevention. The rickettsial diseases generally are diagnosed serologically with complement-fixation and direct immunofluorescence tests.

Concern for the zoonotic potential of these diseases in the laboratory should focus on situations where wild-caught rodents or other small mammals are brought into the laboratory for study or where feral-rodent infestation has occurred. Ectoparasite control in such populations is essential, particularly the elimination of *Ornithonyssus bacoti*, a free-living mite capable of transmitting some of the rickettsial agents (Fox and others 1984). Personnel who are conducting studies with wild-caught animals also should be instructed to practice good laboratory safety and personal hygiene.

BACTERIAL DISEASES

Tuberculosis

Reservoir and Incidence. Tuberculosis of animals and humans is caused by acid-fast bacilli of the genus *Mycobacterium*. Laboratory animals are potential reservoirs of several mycobacterial species, including *M. tuberculosis*, *M. avium-intracellulare*, *M. bovis*, *M. kansasii*, *M. simiae*, *M. marinum*, and *M. chelonae* (Des Prez and Heim 1990; Saunders and Horowitz 1990). In addition to cattle, birds, and humans that serve as the main reservoirs for these mycobacteria, many laboratory animals—including nonhuman primates, swine, sheep, goats, rabbits, cats, dogs, and ferrets—are susceptible to infection and contribute to spread of the diseases (Fox and Lipman 1991). However, nonhuman primates are of primary importance in the consideration of these diseases in the laboratory-animal environment.

Contact with nonhuman primates infected with *Mycobacterium* spp. is a recognized risk factor in the development of a positive tuberculin skin reaction (Kaufman and others 1972). Nonhuman primates generally develop tuberculosis from humans during capture and exportation from parts of the world where the prevalence of the disease in humans and animals is high. However, the resurgence of human tuberculosis in the United States and the recognition of nosocomial outbreaks of multiple-drug-resistant tuberculosis (CDC 1994a) should serve as reminders that nonhuman primates can continue to be at risk for contracting tuberculosis from humans after introduction into established research colonies. The close confinement of these animals in holding facilities and in shipment crates creates an environment conducive to the spread of infection. The incidence of infection in a population varies with the species and the source of the primates. A recent survey of tuberculosis in 22,913 imported nonhuman primates in the United States yielded an incidence of 0.4% (CDC 1993c). Although macaques are considered to be particularly sensitive to infection with *M. tuberculosis*, surveillance programs for tuberculosis should be extended to all species of nonhuman primates (Bennett and others 1995; CDC 1993c; NRC 1980).

Mode of Transmission. *M. tuberculosis* is transmitted via aerosols from infected animals or tissues, and this mode of transmission also applies to most of the other mycobacterial species that might be encountered in laboratory-animal contact. Laboratory personnel involved in the care, use, or necropsy of infected animals are especially at risk for tuberculosis. Humans can contract the disease in the laboratory through exposure to infectious aerosols generated by the handling of dirty bedding, the use of high-pressure water sanitizers, or the coughing of animals with respiratory involvement. Other potential sources of exposure include fecal shedding by animals with enteric infection and skin exudates resulting from scrofuloderma or suppurative fistulated lymph nodes. Mycobacterial disease also can be spread by entry of the bacilli into the body by ingestion or wound contamination.

Clinical Signs, Susceptibility, and Resistance. The most common form of tuberculosis reflects the involvement of the pulmonary system and is characterized by cough, sputum production, and eventually hemoptysis. The incubation period for the development of a demonstrable primary lesion or a substantial secondary skin reaction is 4-12 wk. After that, the risk of progressive pulmonary or extrapulmonary disease remains highest during the next 1-2 yr, but recrudescence of a latent infection persists for the rest of a person's life. Extrapulmonary forms of the disease can involve any tissue or organ system and include disseminated (miliary) infections of multiple organs due to the hematogenous spread of the organism, regional lymphadenitis, tuberculous meningitis, and disease of the pericardium, pleura, skeleton, intestines, peritoneum, kidneys, and skin. General symptoms as the disease progresses include weight loss, fatigue, lassitude, fever, chills, and cachexia.

Diagnosis and Prevention. The diagnosis of tuberculosis in humans and nonhuman primates relies primarily on the use of the intradermal tuberculin test, chest radiography, and the demonstration of acid-fast bacilli in sputum smears. Definitive diagnosis can be obtained by isolating organisms in body fluids or biopsy specimens and identifying them with biochemical techniques or DNA probes. Additional information can be found in guidelines established for the diagnosis and control of tuberculosis in humans (American Thoracic Society 1992; CDC 1994a); revisions have been proposed recently.

The prevention and control of tuberculosis in a biomedical-research facility require personnel education, periodic surveillance for infection in nonhuman primates and their handlers, isolation and quarantine of any suspect animals and prompt euthanasia, necropsy, and microbiological and histopathological analysis of animals confirmed as positive. For extremely valuable animals, chemoprophylaxis with effective antituberculosis agents may be elected (Wolf and others 1988).

The CDC and NIH recommend Animal Biosafety Level 3 for animal studies using nonhuman primates experimentally or naturally infected with *M. tuberculosis* or *M. bovis*. Experimentally infected guinea pigs and mice pose a lesser risk to personnel because droplet nuclei are not produced by coughing in these species; however it is prudent to use Animal Biosafety Level 3 for these infected laboratory animals because contaminated litter can be a source of infectious aerosols (CDC-NIH 1993).

The vaccination of nonhuman primates with the bacillus Calmette Guérin (BCG) strain of *M. bovis* also can be considered. However, the use of BCG does not prevent infection but only suppresses proliferation of the organism to prevent the development of clinical disease (Sutherland and Lindgren 1979). Furthermore, this vaccination complicates the use of the tuberculin test for surveillance because those vaccinated become skin-test-positive. Institutions should consider the implications of BCG vaccination as related to disease monitoring and management in nonhuman primates and the assignment of personnel to the care of these species. Personnel who convert to a positive tuberculin skin reaction should be evaluated further. Institutions should recognize the risk that such personnel pose for nonhuman-primate populations; it might warrant their reassignment to work with other animals. Consistent institutional policies should be developed to address this issue.

Psittacosis (Ornithosis, Parrot Fever, Chlamydiosis)

Reservoir and Incidence. The genus *Chlamydia* contains three species: *C. psittaci*, *C. trachomatis*, and *C. pneumoniae*. Only *C. psittaci* is widely distributed among animals and is recognized as a zoonotic pathogen. *C. psittaci* is distributed widely among birds and mammals worldwide and occurs naturally among many laboratory species, including birds, mice, guinea pigs, rabbits, ruminants, swine, cats, ferrets, muskrats, and frogs (Fox and others 1984; Storz 1971).

Mode of Transmission. C. psittaci produces a diverse spectrum of conditions in animals, including conjunctivitis, pneumonitis, air sacculitis, pericarditis, hepatitis, enteritis, arthritis, meningoencephalitis, urethritis, endometritis, and abortion. Latency is a common characteristic of the infections and is especially important in the epizootology of the disease in birds; stress can reactivate enteric shedding of the organism and clinical signs. The organism is spread to humans from infectious material in exudates, secretions, or desiccated fecal material via direct contact or the aerosol route.

Clinical Signs, Susceptibility, and Resistance. In general, the *C. psittaci* strains associated with mammalian infections are less pathogenic for humans than the avian strains of the organism (Schachter and Dawson 1978). Human conjunctivitis has been observed in people involved in the care of cats with chlamydial conjunctivitis and pneumonitis (Schachter and others 1969). Human abortion resulting from infection with a *C. psittaci* strain that is associated with abortions in sheep also has been recorded (Hadley and others 1992).

The progression of disease in humans related to infection with avian strains of *C. psittaci* includes fever, headache, myalgia, chills, and upper or lower respiratory tract disease. More serious manifestations of disease also can occur, such as extensive pneumonia, hepatitis, myocarditis, thrombophlebitis, and encephalitis. Relapses occur in untreated infections (Benenson 1995b).

Diagnosis and Prevention. Psittacosis can be diagnosed with serological tests for specific antibody or isolation of the organism.

Psittacosis can be prevented by permitting birds only from disease-free flocks to be introduced into an animal facility. If wild-caught birds or birds of unknown disease status are brought into a facility, chlortetracycline chemoprophylaxis should be instituted in these birds. Cases of chlamydiosis in other animals should be treated promptly to prevent the spread of infection to personnel who work with them.

Animal Biosafety Level 2 practices, containment equipment and facilities, and respiratory protection are recommended for personnel working with naturally or experimentally infected caged birds (CDC-NIH 1993).

Rat-Bite Fever

Reservoir and Incidence. Rat-bite fever is caused by either *Streptobacillus moniliformis* or *Spirillum minor*, two microorganisms that are present in the upper respiratory tracts and oral cavities of asymptomatic rodents, especially rats (Anderson and others 1983). These organisms are present worldwide in rodent populations,. although efforts by commercial suppliers of laboratory rodents to eliminate *Strep. moniliformis* from their rodent colonies now appear to have been largely successful. The form of the disease caused by *Spir. minor* can be differentiated clinically from the form due to *Strep. moniliformis* and is generally more common in Asia. Several cases of the disease in laboratory-animal handlers have been reported in recent years (Anderson and others 1983; Taylor and others 1984).

Mode of Transmission. Most human cases result from a bite wound inoculated with nasopharyngeal secretions, but sporadic cases have occurred without a history of rat bite. Infection also has been transmitted via blood of an experimental animal. Persons working or living in rat-infested areas have become infected even without direct contact with rats (Benenson 1995b).

Clinical Signs, Susceptibility, and Resistance. In *Strep. moniliformis* infections, patients develop chills, fever, malaise, headache, and muscle pain and then a maculopapular or petechial rash most evident on the extremities. Arthritis occurs in 50% of *Strep. moniliformis* cases but is considered rare in *Spir. minor* infections. One or more large joints usually become painful and enlarged and contain a serous to purulent effusion. Complications of untreated cases of the disease include focal abscesses, endocarditis, and, less frequently, pneumonia, hepatitis, pyelonephritis, and enteritis.

Diagnosis and Prevention. The disease is diagnosed by isolating the causative organisms, both of which have unusual growth requirements (Fox and others 1984). *Strep. moniliformis* can be isolated in vitro from joint fluid, but *Spir. minor* requires animal inoculation and identification of the organism with dark-field microscopy.

Proper animal-handling techniques are critical to the prevention of rat-bite fever.

Plague

Reservoir and Incidence. Plague, caused by *Yersinia pestis*, has never been recognized as an important disease entity in the laboratory-animal setting. However, focal outbreaks of this once-devastating disease continue to be recognized worldwide, including in the United States, where the disease exists in wild rodents in the western one-third of the country. In the United States, most human cases are related to wild rodents, but cats, dogs, coyotes, rabbits, and goats have also been associated with human infection (Rollag and others 1981; Rosner 1987).

Mode of Transmission. Most human cases are the result of bites by infected fleas or contact with infected rodents. In human plague associated with nonrodent species, infection has resulted from bites or scratches, handling of infected animals (especially cats with pneumonic disease), ingestion of infected tissues, and contact with infected tissues. Nonrodent species can serve as transporters of fleas from infected rodents into the laboratory (Fox and others 1984).

Clinical Signs, Susceptibility, and Resistance. Human plague has a localized (bubonic) form and a septicemic form. In bubonic plague, patients have fever and large, swollen, inflamed, and tender lymph nodes, which can suppurate. The bubonic form can progress to septicemic plague with dissemination of the organism to diverse parts of the body, including the lungs and meninges. The development of secondary pneumonic plague is of special importance because aerosol droplets can serve as a source of primary pneumonic or pharyngeal plague, creating a potential for epidemic disease.

Diagnosis and Prevention. Many tests are used for early rapid diagnosis of plague, including direct microscopic examination of clinical specimens, a fluorescent-antibody (FA) test of tissue specimens, and an antigen-capture ELISA test. Diagnosis is confirmed by culture and identification of the organism or demonstration of a change in antibody titer by a factor of 4 or more (Benenson 1995b).

Preventive measures in a laboratory-animal facility should encompass the control of wild rodents and the quarantine, examination, and ectoparasite treatment of incoming animals with potential infection. Those measures need to be applied continuously for animals that are housed outdoors and therefore have an opportunity for contact with plague-infected animals or their fleas. Vaccines are available for personnel in high-risk categories but confer only brief immunity (Benenson 1995b).

Animal Biosafety Level 2 practices, containment equipment and facilities are recommended for personnel working with naturally or experimentally infected animals (CDC-NIH 1993).

Brucellosis

Reservoir and Incidence. The incidence of brucellosis, which is caused by *Brucella* spp., in agricultural species in the United States is low because eradication of the disease is emphasized. Foci of infection persist in cattle, swine, and ruminant populations. Although zoonotic transmission of the disease from those species is not considered important in the laboratory, *B. suis* of swine might achieve importance as the use of swine in the laboratory increases. *B. canis* in dogs remains a zoonotic hazard in the laboratory-animal facility; canine brucellosis has been identified in dog-production colonies and in 1-6% of dog populations, depending on the geographic area sampled (Fox and others 1984).

Mode of Transmission. Most of the reported human cases of *B. canis* infection have resulted from contact with aborting bitches, and placental tissues are typically rich in organisms in infected animals. *B. canis* also produces prolonged bacteremia and can be present in the urine of infected animals (Mumford and others 1975). Direct contact with the skin or mucous membranes during specimen handling or preparation in the laboratory has resulted in transmission; aerosol transmission also has resulted in large outbreaks of the disease in the laboratory setting. The portal of entry is less well defined in animal-associated transmission of the disease, and a low incidence of seroconversion after exposure might indicate a low likelihood of transmission of the disease.

Clinical Signs, Susceptibility, and Resistance. Human infection with *B. canis* is characterized by fever, headache, chills, myalgia, nausea, and weight loss. Bacteremia can occur, and other systemic involvement is manifested by generalized lymphadenopathy and splenomegaly. Subclinical and inapparent infections also can occur (Benenson 1995b), as evidenced by the seroconversion of 0.5% of asymptomatic military personnel who had contact with infected dogs (Mumford and others 1975).

Diagnosis and Prevention. Organism isolation and serological tests that show a rise in antibody titer are the principal means of diagnosis. Preventive measures should be aimed at excluding infected animals from the facility. Animal handlers should wear appropriate protective clothing and practice good personal hygiene to prevent transmission.

Animal Biosafety Level 3 practices, containment equipment and facilities are recommended for animal studies involving *B. canis*, *B. abortus*, *B. melitensis*, or *B. suis* (CDC-NIH 1993).

Leptospirosis

Reservoir and Incidence. Leptospirosis has a worldwide distribution in domestic and wild animals. Rats, mice, field moles, hedgehogs, squirrels, gerbils, hamsters, rabbits, dogs, domestic livestock, other mammals, amphibians, and reptiles are among the animals that are considered reservoir hosts (Benenson 1995b; Hanson 1982). Pathogenic leptospires belong to the species *Leptospirosis interrogans* and are divided into serovars according to serological reactivity. In the United States, the predominant serovars are *L. icterohaemorrhagia* (in rats and dogs), *L. pomona* (in swine), *L. hardjo* (in cattle), *L. canicola* (in dogs), *L. autumnalis* (in raccoons), and *L. bratislava* (in swine). Rats and mice are common hosts of *L. ballum*, which also has been found in other wildlife, including skunks, rabbits, opossums, and wild cats (Fox and others 1984). The possibility of zoonotic transmission of leptospirosis from most animal species maintained in the laboratory would have to be considered. Several recent outbreaks of the disease in laboratory animals emphasize the continued importance of this zoonosis in the laboratory-animal facility (Alexander 1984; Barkin and others 1974; Geller 1979).

Mode of Transmission. Leptospires are shed in the urine of reservoir animals, which often remain asymptomatic and carry the organism in their renal tubules for years. Mice infected with *L. ballum* are believed to harbor the organism for life (Fox and others 1984). Transmission occurs through skin abrasions and mucous membranes and is often related to direct contact with urine or tissues of infected animals. Inhalation of infectious droplet aerosols and ingestion also are effective modes of transmission.

Clinical Signs, Susceptibility, and Resistance. The manifestations of this disease are diverse, ranging from inapparent infection to severe systemic illness (Benenson 1995b). Common features are fever with sudden onset, headache, chills, myalgia, and conjunctival suffusion. Other manifestations of the disease are orchitis, rash, hemorrhage into the skin and mucous membranes, hemolytic anemia, hepatorenal failure and jaundice, mental confusion with encephalitis, and pulmonary involvement.

Diagnosis and Prevention. Leptospirosis is diagnosed by showing rising antibody titers in serological tests, such as the microscopic agglutination test, or by isolating the organism. Efforts to prevent this zoonotic disease in a laboratory-animal facility should focus on effective control of the infection in laboratory-animal populations and use of protective clothing and gloves by personnel.

Campylobacteriosis

Reservoir and Incidence. Organisms of the genus *Campylobacter* have been recognized as a leading cause of diarrhea in humans and animals in recent years, and numerous cases involving the zoonotic transmission of the organisms in pet and laboratory animals have been described (Blaser and others 1980; Deming and others 1987; Fox 1982; Fox and others 1989a,b; Russell and others 1990). Results of prevalence studies on dogs, cats, nonhuman primates, and group-housed animals suggest that young animals readily acquire the infection and shed the organism; young animals often are implicated as the source of infection in zoonotic transmission.

Mode of Transmission. The organism is transmitted by the fecal-oral route via contaminated food or water or direct contact with infected animals.

Clinical Signs, Susceptibility, and Resistance. Campylobacters produce an acute gastrointestinal illness, which in most cases is brief and self-limiting. The clinical signs of campylobacter enteritis include watery diarrhea, sometimes with mucus, blood, and leukocytes; abdominal pain; fever; and nausea and vomiting. The infection generally resolves with specific antimicrobial therapy. Unusual complications of the disease include a typhoid-like syndrome, reactive arthritis, hepatitis, interstitial nephritis, hemolytic-uremic syndrome, febrile convulsions, meningitis, and Guillain-Barré syndrome (Benenson 1995b, Blaser 1990).

Diagnosis and Prevention. Organism isolation is used to diagnose campylobacter infection. Although the treatment of animals with campylobacter enteritis is useful in the control of the infection, the attempt to eliminate the carrier state in asymptomatic animals might be less rewarding. Personnel should rely on the use of protective clothing, personal hygiene, and sanitation measures to prevent the transmission of the disease.

Animal Biosafety Level 2 is recommended for activities using naturally or experimentally infected animals (CDC-NIH 1993).

Salmonellosis

Reservoir and Incidence. Enteric infection with *Salmonella* spp. has a worldwide distribution among humans and animals. Among the laboratory-animal species, rodents from many sources are now free from salmonella infection because of successful programs of cesarean rederivation accompanied by rigorous management practices to exclude the recontamination of animal colonies. The pasteurization of feeds also has contributed to the control of salmonellae in laboratory-animal populations. However, despite those efforts to eliminate the organisms in laboratory-animal populations, salmonella carriers continue to occur as a result of infection by contaminated food or other environmental sources of contamination and represent a source of infection for other animals and personnel who work with the animals (Nicklas 1987).

Results of recent surveys in dogs and cats have indicated that the prevalence of infection remains about 10% among random-source animals (Fox and Lipman 1991). Salmonellae continue to be recorded frequently among recently imported nonhuman primates (Tribe and Fleming 1983). Infection with salmonellae is nearly ubiquitous among reptiles; during the 1970s, salmonellosis in turtles was a major public-health concern, which was eventually controlled by restricting the sale of viable turtle eggs or live turtles with a carapace length of at least 10.2 cm to institutions with a scientific or educational mission. Avian sources are often implicated in foodborne cases of human salmonellosis, and birds should be considered likely sources of zoonotic transmission in a laboratory-animal facility.

Mode of Transmission. Salmonellae are transmitted by the fecal-oral route via food derived from infected animals or contaminated during preparation, contaminated water, or direct contact with infected animals.

Clinical Signs, Susceptibility, and Resistance. Salmonella infection produces an acute febrile enterocolitis; septicemia and focal infections occur as secondary complications (Benenson 1995b; Hook 1990). Focal infections can be localized in any tissue of the body, so the disease has diverse manifestations. Many host factors have been associated with increased severity of the disease, including infancy, old age, AIDS, neoplasia, immunosuppressive therapy or other debilitating condition, achlorhydria, gastrointestinal surgery, or prior or current broad-spectrum antibiotic therapy.

Diagnosis and Prevention. Organism isolation with standard microbiological techniques is used to diagnose this infection. Concomitant isolation of the same organism as determined with appropriate molecular biology and molecular epidemiology can be used to implicate a suspect animal as a source of zoonotic transmission.

Whenever possible, animals known not to harbor salmonellae should be used in laboratory-animal facilities, and the combination of microbiological screening of individual animals or a representative sample of the animal population for the presence of salmonellae and isolation or elimination of carriers can aid in excluding the pathogen from an animal facility. The use of antibiotic treatment of salmonella-infected animals as a means of controlling the organism in a laboratory-animal facility might not be rewarding, because antibiotic treatment can prolong the period of communicability (Benenson 1995b). Personnel should rely on the use of protective clothing, personal hygiene, and sanitation measures to prevent the transmission of the disease.

Animal Biosafety Level 2 is recommended for activities using naturally or experimentally infected animals (CDC-NIH 1993).

Shigellosis

Reservoir and Incidence. Nonhuman primates are the only important reservoir for shigella infection in animal facilities (Fox and others 1984; Richter and others 1984), although zoonotic transmission of the organism from guinea pigs, other rodents, and dogs has been recorded under unique circumstances (Benenson 1995b; CDC-NIH 1993). Nonhuman primates can harbor several *Shigella* spp. that are pathogenic for humans, including *S. flexneri, S. sonnei,* and *S. dysenteriae.* The organisms produce in nonhuman primates a diarrheal disease similar to that seen in humans. Nonhuman-primate infections occur as a result of contact with other infected primates, including humans, or contaminated food, water, or fomites.

Mode of Transmission. Shigellosis is transmitted by a direct or indirect fecal-oral route. *Shigella* spp. are extremely infectious, requiring only 10-100 organisms to produce infection.

Clinical Signs, Susceptibility, and Resistance. Shigellosis is characterized by an acute onset of diarrhea accompanied by fever, nausea and sometimes vomiting, tenesmus, cramps, and toxemia (Benenson 1995b). In contrast with findings in salmonellosis, bacteremia is very uncommon. The diarrhea is often watery, containing blood, mucus, and pus; and it can be life-threatening in the elderly, debilitated, and malnourished. All age groups are susceptible to infection, but healthy adults infected with a small number of organisms can develop asymptomatic infection.

Diagnosis and Prevention. Routine microbiological methods are used to isolate and identify shigellae. The prevention of shigellosis in a laboratory-animal facility should be based on identification and treatment of the carrier state or disease in a nonhuman-primate reservoir (Fox and Lipman 1991). Personnel also should rely on the use of protective clothing, personal hygiene, and sanitation measures to prevent the transmission of the disease.

Animal Biosafety Level 2 is recommended for activities using naturally or experimentally infected animals (CDC-NIH 1993).

Enteric Yersiniosis

Reservoir and Incidence. Yersinia enterocolitica and *Y. pseudotuberculosis* are present in a wide variety of wild and domestic animals, which are considered the natural reservoirs for the organisms. The host species for *Y. enterocolitica* include rodents, rabbits, pigs, sheep, cattle, horses, dogs, and cats; *Y. pseudotuberculosis* has a similar host spectrum and also includes various avian species (Butler 1990). Human infections often have been associated with household pets, particularly sick puppies and kittens (Benenson 1995b). Occasional reports of yersinia infections in animals housed in the laboratory—such as guinea pigs, rabbits, and nonhuman primates—suggest that zoonotic yersinia infection should not be overlooked in this environment (Fox and others 1984).

Mode of Transmission. Yersinia spp. are transmitted by direct contact with infected animals through the fecal-oral route.

Clinical Signs, Susceptibility, and Resistance. *Y. enterocolitica* produces a gastroenterocolitis syndrome characterized by fever, diarrhea, and abdominal pain. In some cases, ulcerative mucosal lesions occur in the terminal ileum; they are often accompanied by mesenteric lymphadenitis mimicking the clinical presentation of acute appendicitis (Butler 1990). Other serious sequelae of infection include postinfectious arthritis, iritis, skin ulceration, hepatosplenic abscesses, osteomyelitis, and septicemia.

Diagnosis and Prevention. Most clinically important infections can be detected with routine enteric culturing methods, although cold enrichment, alkali treatment, or selective CIN agar can be used to enhance growth of the organisms. Laboratory animals with yersiniosis should be isolated and treated or culled from the colony. Personnel should rely on the use of protective clothing, personal hygiene, and sanitation measures to prevent the transmission of the disease.

PROTOZOAL DISEASES

Vector-borne protozoal diseases generally are not considered a direct threat to personnel in laboratories, because the importation of vectors with hosts is highly improbable. However, accidental inoculation and wound contamination with infected tissue derivatives are conceivable means of transmitting plasmodal, trypanosomal, and leishmanial infections, and appropriate precautions should be taken by personnel who work with these agents in animals.

Toxoplasmosis

Reservoir and Incidence. *Toxoplasma gondii* is a coccidian parasite with a worldwide distribution among warm-blooded animals. Wild and domestic felines are the only definitive hosts of this organism; they are infected by one another or through predation of an intermediate host, and they support all phases of the *T. gondii* life cycle in their intestinal tract, although numerous other tissues are also involved in feline toxoplasmosis (Dubey and Carpenter 1993). Results of serological surveys have indicated that 30-80% of cats have evidence of *T. gondii* infection (Ladiges and others 1982). Intermediate hosts, including humans, can contract the infection from oocysts, which are present only in materials contaminated by cat feces, or by ingesting infectious bradyzoites or cystozoites encysted in the tissues of another infected animal. In a laboratory-animal facility, the control of this zoonosis is centered principally around the management of cats (Fox and others 1984). Although many other laboratory animals could serve as intermediate hosts and harbor *T. gondii* in extraintestinal sites, they have not proved to be important sources of zoonotic transmission in the laboratory environment.

Mode of Transmission. Infection results from the ingestion of infectious oocysts in food, water, or other sources contaminated by feline feces. The ingestion of uncooked or undercooked meat, especially pork and beef, is an important source of human infection. Consequently, human infection from improper handling of tissue of an infected intermediate host in the laboratory should be considered a remote possibility.

Clinical Signs, Susceptibility, and Resistance. Toxoplasmosis generally produces an asymptomatic or mild infection with fever, myalgia, arthralgia, lymphadenopathy, and hepatitis (Benenson 1995b). Toxoplasma infection can have severe consequences in pregnant women and immunologically impaired people. In a pregnant woman with a primary infection, rapidly dividing tachyzoites can circulate in the bloodstream and produce a transplacental infection of the fetus. In early pregnancy, the fetal infection can result in death of the fetus or chorioretinitis, brain damage, fever, jaundice, rash, hepatosplenomegaly, and convulsions at birth or shortly thereafter. Fetal infection during late gestation can result in mild or subclinical disease with delayed manifestations, such as recurrent or chronic chorioretinitis. Primary infection in immunosuppressed people can be characterized by maculopapular rash, pneumonia, skeletal myopathy, myocarditis, brain involvement, and death.

Diagnosis and Prevention. Toxoplasmosis can be diagnosed by finding the organism in clinical specimens, isolating it in an animal or cell culture, or demonstrating rising antibody titers.

Personnel should practice appropriate personal-hygiene practices and maintain rigorous sanitation of an animal facility to prevent exposure to toxoplasma. Unless they are known to have antibodies to toxoplasma, pregnant women should be advised of the risk associated with fetal infection. Cat feces and litter should be disposed of promptly before sporocysts become infectious, and gloves should be worn in the handling of potentially infective material.

Giardiasis

Reservoir and Incidence. Many wild and laboratory animals serve as a reservoir for *Giardia* spp., although cysts from human sources are regarded as more infectious for humans than are those from animal sources (Benenson 1995c). Dogs, cats, and nonhuman primates are the laboratory animals most likely to be involved in zoonotic transmission. According to recent surveys of endoparasites in dogs, the prevalence of giardia generally ranges from 4 to 10% and approaches 100% in some breeding kennels (Jordan and others 1993; Kirkpatrick 1990).

Mode of Transmission. Giardiasis is transmitted by the fecal-oral route chiefly via cysts from an infected person or animal. The organism resides in the upper gastrointestinal tract where trophozoites feed and develop into infective cysts.

Clinical Signs, Susceptibility, and Resistance. Humans and animals have similar patterns of infection. Infection can be asymptomatic, but anorexia, nausea, abdominal cramps, bloating, and chronic, intermittent diarrhea are often seen. Although the organism is rarely invasive, severe infections can produce inflammation in the bile and pancreatic ducts and damage the duodenal and jejunal mucosa, resulting in the malabsorption of fat and fat-soluble vitamins.

Diagnosis and Prevention. Giardiasis is diagnosed by finding cysts or trophozoites in stool specimens or in duodenal aspirates of humans or animals. Identification and treatment of giardiasis in a laboratory-animal host in combination with effective personal-hygiene measures should reduce the potential for zoonotic transmission in a laboratory-animal facility.

Cryptosporidiosis

Reservoir and Incidence. Cryptosporidium spp. have a cosmopolitan distribution and have been found in many animal species, including mammals, birds, reptiles, and fishes (Fayer and Ungar 1986). Cross-infectivity studies have shown a lack of host specificity for many of the organisms (Tzipori 1988). Among the laboratory animals, lambs, calves, pigs, rabbits, guinea pigs, mice, dogs, cats, and nonhuman primates can be infected with the organisms. Cryptosporidiosis is common in young animals, particularly ruminants and piglets.

Mode of Transmission. Cryptosporidiosis is transmitted by the fecal-oral route and can involve contaminated water, food, and possibly air (Soave and Weikel

1990). Many human cases involve human-to-human transmission or possibly the reactivation of subclinical infections. Several outbreaks of the disease have been associated with surface-water contamination; a recent waterborne epidemic in Milwaukee, Wisconsin, was believed to involve more than 370,000 people (Dresezen 1993). Zoonotic transmission of the disease to animal handlers has been recorded, including a recent report of cryptosporidiosis among handlers of infected infant nonhuman primates; this emphasizes the importance of this zoonosis in the laboratory-animal environment (Anderson 1982; Miller and others 1990; Reese and others 1982).

Clinical Signs, Susceptibility, and Resistance. Although cryptosporidiosis has become identified widely with immunosuppressed people, particularly AIDS patients, the ability of the organism to infect immunocompetent people also has been recognized. In humans, the disease is characterized by cramping, abdominal pain, profuse watery diarrhea, anorexia, weight loss, and malaise (Soave and Weikel 1990). Symptoms can wax and wane for up to 30 d, eventually resolving in immunocompetence. However, in AIDS patients, who might have an impaired ability to clear the parasite, the disease can have a prolonged course that contributes to death.

Diagnosis and Prevention. Cryptosporidiosis is diagnosed by finding the organism in stool specimens with immunofluorescent or other special staining techniques (Soave and Weikel 1990). Several samples might be necessary because of intermittent shedding of the organism. Appropriate personal-hygiene practices should be effective in preventing the spread of infection. No pharmacological treatment is effective for this infection.

Amebiasis

Reservoir and Incidence. Humans serve as the reservoir for *Entamoeba histolytica*, the causative agent of amebiasis, although nonhuman-primate infections have been recorded (Fox and others 1984). The importance of nonhuman primates as a reservoir host appears to have diminished in recent years.

Mode of Transmission. The disease is transmitted by ingestion of amebic cysts that are present in the feces of infected animals.

Clinical Signs, Susceptibility, and Resistance. Clinical signs of amebiasis can range from mild abdominal discomfort with intermittent diarrhea containing blood and mucus to acute fulminating dysentery with fever, chills, and bloody or mucoid diarrhea. In severe cases, the organism can penetrate the colonic mucosa, become disseminated in the bloodstream, and produce liver, lung, or brain abscesses.

Diagnosis and Prevention. The disease is diagnosed by finding cysts or trophozoites in fresh fecal specimens or other clinical specimens. Nonhuman-primate carriers of the infection should be identified and treated. Appropriate facility sanitation and personal-hygiene practices should prevent the zoonotic transmission of the agent.

Balantidiasis

Reservoir and Incidence. Balantidium coli has a worldwide distribution and is common in domestic swine, which generally are regarded as the main reservoir for human infection. Nonhuman primates also can harbor the organism enterically (Fox and others 1984).

Mode of Transmission. The agent is transmitted by the fecal-oral route.

Clinical Signs, Susceptibility, and Resistance. Most humans appear to have a high natural resistance to this infection. However, ulcerative colitis characterized by diarrhea, abdominal pain, tenesmus, nausea, and vomiting can occur in severe cases of the disease.

Diagnosis and Prevention. The treatment of clinically apparent infections in a laboratory-animal host should be coupled with good sanitation and personal-hygiene practices to eliminate the zoonotic transmission of this organism in an animal facility.

FUNGAL DISEASES

Dermatomycosis

Reservoir and Incidence. The dermatophytes have a cosmopolitan distribution; some dermatophytes have a regional geographic concentration (Benenson 1995b). These organisms cause ringworm in humans and animals, which continues to be common among dogs, cats, and livestock (Fox and others 1984). In the United States, several dermatophytes of animal origin are involved in the superficial mycoses of humans, including *Microsporum canis, Trichophyton mentagrophytes,* and *T. verrucosum. M. canis* is most prevalent in dogs, cats, and nonhuman primates and in human infections associated with these species, but it can also occur in rodents. *T. mentagrophytes* has been associated more commonly with ringworm in rodents and rabbits and occurs among laboratory personnel who work with these species and agricultural personnel who work around granaries, barns, and other rodent habitats. *T. verrucosum* is restricted generally to cases of ringworm in livestock and their agricultural attendants.

Mode of Transmission. The transmission of dermatophyte infection from humans to animals is by direct skin-to-skin contact with infected animals or indirect contact with contaminated equipment or materials. Infected animals can have no, few, or difficult-to-detect skin lesions that result in transmission to unsuspecting persons. Dermatophyte spores can become widely disseminated and persistent in the environment, contaminating bedding, equipment, dust, surfaces, and air and resulting in the infection of personnel who do not have direct animal contact.

Clinical Signs, Susceptibility, and Resistance. The clinical expression of dermatomycosis depends on various host factors and the predilection of the organism. Dermatophytes generally grow in keratinized epithelium, hair, nails, horn, and feathers and are classified according to their optimal substrate as geophilic (soil), zoophilic (animals), or anthropophilic (human). Many of the zoophilic fungi are species-adapted and cause infection without inciting serious inflammatory lesions in their host species; however, in an aberrant host, such as a human, a vesicular or pustular eczematous lesion with intense inflammation and rapid regression can occur. Dermatophytes that are better adapted to humans produce focal, flat, spreading annular lesions that are clear in the center and crusted, scaly, and erythematous in the periphery. Lesions often are on the hands, arms, or other exposed areas, but invasive and systemic infections have been reported in immunocompromised people.

Diagnosis and Prevention. The definitive diagnosis of dermatomycosis is achieved by fungal culture and identification, but lesion appearance and scrapings of active lesions cleared in 10% potassium hydroxide and examined microscopically for fungal filaments can be used for a tentative diagnosis. In addition, about half of *M. canis* isolates and lesions are fluorescent in Wood's lamp examination.

Animals with suggestive lesions should be screened for dermatomycosis and isolated and treated if positive. The use of protective clothing, disposable gloves, and other appropriate personal-hygiene measures is essential to the reduction of this zoonosis in a laboratory-animal facility.

Animal Biosafety Level 2 practices and facilities are recommended for experimental animal activities with dermatophytes (CDC-NIH 1993).

Sporotrichosis

Reservoir and Incidence. Sporothrix schenckii is a fungal agent reported in all parts of the world and generally associated with agricultural occupations. However, sporotrichosis has been reported in numerous laboratory-animal species, including dogs, cats, swine, cows, goats, rats, and armadillos (Werner and Werner 1993).

Mode of Transmission. Most cases of zoonotic transmission have implicated the direct inoculation of the fungus into bites or skin wounds inflicted by animals, but several people who have developed infections could not recall pre-existing skin lesions or skin injury in conjunction with exposure. Thus, this organism might be capable of penetrating intact skin.

Clinical Signs, Susceptibility, and Resistance. Humans usually develop a solitary nodule on the hand or extremity and nodular extension along the path of the lymphatic vessels. Ulceration and drainage of the lesions can occur. Arthritis, pneumonia, and other deep visceral infections occur as rare complications (Benenson 1995b).

Diagnosis and Prevention. Sporotrichosis is diagnosed by culture and identification of the organism with Sabouraud dextrose agar. Animals with known or suspected sporothrix infections should be isolated and treated, and personnel should practice appropriate personal-hygiene measures when handling these animals.

Animal Biosafety Level 2 practices and facilities are recommended for activities using naturally or experimentally infected animals (CDC-NIH 1993).

HELMINTH INFECTIONS

Despite the large number of helminth-parasite infections that either are directly zoonotic or have cycles of infection that encompass animals and humans (see Table 5-1), the transmission of helminthic zoonoses in the laboratory-animal environment should be regarded as unlikely (Fox and others 1984). Many of the organisms have indirect life cycles that are interrupted in the laboratory environment or have ova embryonation periods that are long enough to permit removal of ova during routine sanitation before they become infective for humans (Flynn 1973). In addition to contemporary laboratory-animal management practices that impede zoonotic transmission of helminth parasites, animal-health conditioning practices should be in place to eliminate infections. The use of appropriate personal-hygiene practices also must be emphasized to eliminate any possibility of zoonotic infection.

ARTHROPOD INFESTATIONS

Very few ectoparasite infestations of humans are associated with the handling of conventional laboratory animals, but several have been reported (Fox and others 1984). Appropriate attention needs to be given to the control of this risk; animals are introduced from the wild, animals are used in studies under natural field conditions, or conventional laboratory animals are used in facilities whose vermin-control measures are inadequate to preclude the introduction of these agents on endemically infected wild-animal reservoirs.

TABLE 5-1 Zoonotic Helminth Parasites of Laboratory Animals

Zoonosis	Parasite	Host	Comments
Ascariasis	*Ascaris lumbricoides*	Old World primates	Infection occurs by ingestion of embryonated eggs only; embryonation, requiring 2 weeks or more, ordinarily would not occur in laboratory; heavy infections can produce severe respiratory and gastrointestinal tract disease.
Cestodiasis	*Hymenolepis nana*	Rat, mouse, hamster, nonhuman primates	Intermediate host is not essential to life cycle; direct infection and internal autoinfection can occur also; heavy infections result in abdominal distress, enteritis, anal pruritus, anorexia, and headache.
Larval migrans (cutaneous)	*Ancylostoma caninum*	Dog	Transcutaneous infection causes parasitic dermatitis called "creeping eruption."
	Ancylostoma braziliense	Dog, cat	
	Ancylostoma duodenale	Dog, cat	
	Uncinaria stenocephala	Dog, cat	
	Necator americanus	Dog, cat	

Generally, human ectoparasite infestations are manifested as mild allergic dermatitis (see Table 5-2). The more-important, albeit rarer, risk associated with these infestations is transmission of zoonotic agents that can produce systemic disease with arthropods as a vector. Every major group of pathogenic organisms—including bacteria, rickettsiae, chlamydia, viruses, protozoa, spirochetes, and helminths—is represented among the agents transmitted by arthropod vectors, and personnel who work with research animals that potentially harbor these agents or the ectoparasite vectors should be informed of the hazard.

Rigorous ectoparasite-control programs should be instituted as part of the veterinary-care program, especially for wild-caught species that are brought into a laboratory, animals housed previously under field conditions, and animals with inadequate disease profiles from any source. The control of vermin in an animal facility also is essential; consideration should be given to the ectoparasite and disease evaluation of wild or feral rodents caught in an animal facility.

TABLE 5-1 Continued

Zoonosis	Parasite	Host	Comments
Larval migrans (visceral)	*Toxocara canis* *Toxocara cati* *Toxocara leonina*	Dog Cat Dog, cat	Chronic eosinophilic granulomatous lesions distributed throughout various organs; should not be encountered in laboratory.
Strongyloidiasis	*Strongyloides stercoralis*, *Strongyloides fulleborni*	Old World primates, dog, cat	Oral and transcutaneous infections can occur in animals and humans; heavy infections can produce dermatitis, verminous pneumonitis, and enteritis; internal autoinfection can occur.
Oesophagostomiasis	*Oesophagostomum* spp.	Old World primates	Heavy infections result in anemia; encapsulated parasitic granulomas are usually innocuous sequelae of infection.
Ternidens infection	*Ternidens deminutus*	Old World primates	Rare and asymptomatic.
Trichostrongylosis	*Trichostrongylus colubriformis*, *Trichostrongylus axei*	Ruminants, pig, dog, rabbit, Old World primates	Heavy infections produce diarrhea.

Source: Adapted from: Fox and others 1984.

TABLE 5-2 Zoonotic Ectoparasites of Laboratory Animals

Species	Disease in Humans	Host	Comments
Fleas			
Ctenocephalides felis, C. canis	Dermatitis	Dog, cat	Vector of *Hymenolepis diminuta, Dipylidium caninum*
Xenopsylla cheopsis	Dermatitis	Mouse, rat, wild rodents	Vector of *H. nana, H. diminuta*
Nasopsyllus fasciatus	Dermatitis	Mouse, rat, wild rodents	Vector of *H. nana, H. diminuta, R. mooseri*
Leptopsylla segnis	Dermatitis	Rat	Vector of *H. diminuta, H. nana, R. mooseri*
Pulex irritans	Irritation	Domestic animals (especially pig)	
Mites			
Obligate skin mites			
Sarcoptes scabiei subspp.	Scabies	Mammals	
Notoedres cati	Mange	Cat, dog, rabbit	
Nest-inhabiting parasites			
Ornithonyssus bacoti	Dermatitis	Rodents and other vertebrates, including birds	Vector of western equine encephalitis and St. Louis encephalitis viruses, *Rickettsia mooseri*
Allodermanyssus sanguineus	Dermatitis	Rodents, particularly *Mus musculus*	Vector of *Rickettsia akari*
Trixacarus cavae	Dermatitis	Guinea pig	
Facultative mites			
Cheyletiella spp.	Dermatitis	Cat, dog, rabbit (bedding)	

TABLE 5-2 Continued

Species	Disease in Humans	Host	Comments
Ticks			
Rhipicephalus sanguineus	Irritation	Dog	Vector of *Rickettsia rickettsia, Francisella tularensis, Ehrlichia canis*
Dermacentor variabilis	Irritation	Wild rodents, cottontail rabbit, dogs from endemic areas	Vector of *Rickettsia rickettsia, Francisella tularensis, Ehrlichia canis*
Dermacentor andersoni	Irritation	Small mammals, uncommon on dog	Vector of *Rickettsia rickettsia, Francisella tularensis, Ehrlichia canis*
Dermacentor occidentalis	Irritation	Small mammals, uncommon on dog	Vector of *Rickettsia rickettsia, Francisella tularensis, Ehrlichia canis*
Amblyomma americanum	Irritation	Wild rodents, dog	
Ixodes scapularis	Irritation		
Ixodes dammini	Irritation	Dog, wild rodents	Vector of *Borrelia burgdorferi, Babesia microtis*

Adapted from: Fox and others 1984.

6

Principal Elements of an Occupational Health and Safety Program

Most institutions have developed effective programs for controlling hazards and minimizing occupational risks of injury and illness in the workplace. The motivation for and commitment to conducting occupational health and safety programs are derived from two principal sources: a moral obligation to safeguard employees from unnecessary risks and a regulatory requirement that employers provide a safe and healthful workplace for their employees. Many institutions that maintain an animal care and use program have an environmental health and safety office that involves people with expertise in chemical safety, biological safety, physical safety, industrial hygiene, health physics, engineering, environmental health, occupational health, fire safety, and toxicology or have corresponding technical resources available under other arrangements. The environmental health and safety office generally provides technical consultation, risk assessment, accident reviews, training, emergency response, waste management, recordkeeping, inspections and audits, and compliance monitoring. Those services assist institutional leaders and managers of the animal care and use activities in establishing health and safety policies and promoting high standards of safety. Services provided by the environmental health and safety office can be most helpful, however, when they are designed in collaboration with the institutional leaders, managers, and employees so as to ensure that the occupational health and safety program not only complies with regulations but is relevant and practical for the animal care and use program.

There are nine key elements of effective occupational health and safety programs: administrative procedures, facility design and operation, exposure control, education and training, occupational health-care services, equipment performance, information management, emergency procedures, and program evaluation. All but occupational health-care services are discussed in this chapter. The occupational health-care services are often the most difficult for an institution to plan and carry out because consensus on what needs to be done has not yet been established. This important element is discussed separately in Chapter 7.

ADMINISTRATIVE PROCEDURES

Adequate administrative procedures are vital to the success of an occupational health and safety program. Administrative procedures are most effective if developed in collaboration with their users, and both managers and employees need to know their roles. Approval mechanisms established to authorize research activity should be clear, practical, and well publicized.

Procedures should be developed for conducting a health and safety review of research activities that involve infectious agents, recombinant-DNA molecules that are not exempt from federal guidelines, hazardous chemicals, radiation, or the use of animals that present unique hazards. Those procedures should be incorporated into the institutional animal care and use committee (IACUC) project-review process. An appropriate environmental health and safety professional can serve on the committee to participate in the review or be otherwise involved in the review process. Where substantial risks exist, researchers should be encouraged to incorporate health and safety procedures as an integral part of the research plan.

FACILITY DESIGN AND OPERATION

During the design of a new facility or the renovation of an existing one, hazards associated with the care and use of animals should be addressed in a collaborative effort involving investigators who will use the facility, the manager and other principal staff of the institution's animal care and use program, and environmental health and safety staff. The design process begins with defining the species of animals expected to be housed in the facility and the nature of the research programs that will use them. Thorough consideration of hazards is necessary to ensure that the design will allow compliance with federal, state, and local government safety requirements and meet relevant accreditation standards. For example, adequate space should be made available for storage of hazardous materials and for the collection, storage, and processing of wastes. The potential users, the manager of the animal care and use program, a representative of the environmental health and safety staff, the building engineer, and the architect should remain involved in the design and construction process until completion.

Special consideration should be given to the ventilation system, space arrangement and layout, support areas, traffic patterns, and access to utilities and mechanical areas. Criteria for selecting mechanical systems and equipment should be based on reliability, operational integrity, projected length of service, and ease of maintenance. The selection of space, layout of equipment, work surfaces, and traffic patterns will influence the operational effectiveness of the facility and the ease with which staff can maintain established administrative procedures for operating the facility safely. A program of preventive maintenance will ensure continued safe operation of a well-designed facility; this is an important aspect of occupational health and safety, particularly when efforts to minimize substantial risks require the use of engineering controls.

Careful attention should be given to prevention and control of ergonomic hazards in the design of animal facilities (NRC 1996). Engineering controls that reduce physical stress in repetitive operations and in the lifting and movement of heavy loads by animal care staff are important design objectives. Ergonomic design criteria should be used in the selection of fixed equipment, such as animal caging, necropsy tables, and sinks. Several authoritative references are available which provide comprehensive coverage of this important subject (CCAC 1993, DiBerardinis and others 1993, NRC 1996, Ruys 1991).

EXPOSURE CONTROL METHODS

Exposures to occupational hazards are controlled through the application of engineering controls, work practices, and the use of personal protective equipment. Those measures are used in a hierarchical structure. That is, it is first attempted to isolate workers from hazards with engineering controls. If engineering controls do not adequately control the exposure potential, work practices are modified to help to minimize exposure potential. Finally, personal protective equipment might be required to provide a barrier between employees and hazards that cannot be otherwise controlled.

Engineering Controls

Engineering controls are a combination of safety equipment and physical features of the facility that help to minimize hazardous exposures of personnel and the surrounding environment. Safety equipment provides a barrier between employees and hazards, and physical features can prevent or reduce the potential for release of hazardous agents from the immediate work area. Some engineering controls commonly used in animal care and research are barriers and airlocks, chemical fume hoods, biological safety cabinets, and isolation cages.

Barriers help to confine potential contamination to areas where it is generated and to control access to these areas. In animal biosafety level 3 facilities (see Table 3-4), barriers isolate animal areas from other, adjacent areas. The principal barriers are exhaust air ventilation systems that provide directional airflow, architectural barriers that control access to the animal facility, and airlocks that help to maintain air pressure differentials to ensure the proper direction of airflow. Access control barriers also have value for any animal facility because they can be used to prevent unauthorized people from entering the animal facility; this kind of control is difficult to accomplish without constructing an access foyer or special entrance area through which authorized people must pass before entering the facility.

Chemical fume hoods are local exhaust devices that help to prevent toxic, offensive, and flammable vapors or dusts from entering a work area (DiBerardinis and others 1993, NRC 1995). They provide employee protection from such hazards as chemical spills, splashes or sprays, other accidentally released materials, fires, and minor explosions. Hoods should be properly located in the laboratory away from doors, supply air ducts, and high traffic areas. Hoods should be evaluated before use to ensure adequate face velocities (typically 80-100 ft/min) and the absence of excessive turbulence (NRC 1995). The hood installation should include a continuous airflow monitoring device to allow the user to check operating conditions before conducting hazardous procedures. If inadequate hood performance is suspected, correct operation should be verified before the hood is used. The hood sash opening should be kept as narrow as reasonably practicable to improve the overall performance of the hood. The containment capability of hoods is also influenced by the amount and placement of equipment in the hood, persons walking by the hood, and the opening and closing of doors. Careful technique by the user is essential in achieving optimal performance.

Biological safety cabinets are among the most effective, as well as the most commonly used, primary containment devices for work with infectious agents. Several types of cabinets are available, and authoritative references should be reviewed before a cabinet is selected for a particular experimental use (CDC-NIH 1993, Fleming and others 1995, Kruse and others 1991). As with any piece of laboratory equipment, personnel should be trained in the proper use of biological safety cabinets. Air balance and inward airflow are critical in the safe operation of these cabinets. Biological safety cabinets should be certified in accordance with the National Sanitation Foundation Standard 49 (NSF 1992). Containment can be compromised by interruption in airflow caused by insertion and removal of a worker's arms through the work opening, opening and closing of room access doors, and movement of staff near the cabinet. Fans, heating and air conditioning diffusers, and other air-handling devices near the cabinet can also disrupt airflow patterns. Biological safety cabinets have been configured to provide containment space for cleaning cages. They can protect both the animals and personnel from exposures to aerosols that are generated by cleaning procedures.

Cage filter tops are used in animal research to prevent cross contamination with infectious agents. They prevent transmission of agents between and among animals and people by preventing particles from entering the cage. Isolation cages with filter tops that fit tightly to the cage rim can constitute an effective barrier to transmission of agents by the aerosol route, but they should be used in conjunction with a biological safety cabinet to ensure containment during procedures that involve removing the cage top.

Ventilated caging systems also control hazards. Exhaust fans create a negative pressure gradient between the cage and the surrounding environment, and exhaust air is filtered with a high-efficiency-particulate-air (HEPA) filter before discharge into the animal room or the building exhaust; this combination can prevent the escape of bioaerosols from the animal environment.

Downdraft necropsy tables capture chemical vapors generated during necropsy. The tables are constructed with exhaust fans that produce a downdraft by drawing air through the work surface. Air velocities above the work surface, however, are not sufficient to capture aerosols that are generated by the procedure. The protective capacity of these tables can be compromised by air turbulence in the room, the size of the animal on the table, and general work practices. Their use should be carefully assessed by knowledgeable health and safety professionals.

Room ventilation is an important engineering control used to maintain comfortable temperature and humidity in the work area. Changing air continuously can reduce the concentration of airborne contaminants but does not replace the use of such containment devices as chemical fume hoods, biological safety cabinets, and filter top cages. A ventilation system that provides directional airflow can prevent the migration of airborne contaminants to unprotected space in the facility.

Cage cleaning and cage washing can result in high concentrations of particulate contaminants and very high heat loads from the cage washing equipment. Consequently, high ventilation rates are important for providing acceptable environmental conditions for personnel.

Local exhaust can be effective in controlling contaminants at the point of generation. Properly engineered and used canopy hoods and flexible exhaust ducts can substantially reduce occupational exposures to such hazards as animal dander and excreta liberated during cage cleaning, aerosols and vapors generated during anesthesia or necropsy, and heat emanating from cage cleaning or waste decontamination. Slot hoods can also be used in controlling these exposures, but their effectiveness depends on the correct static pressure, flow rate, and hood geometry (NRC 1995, p.# 190). Local exhaust devices are particularly useful for controlling emissions from equipment or procedures that cannot reasonably be contained in a hood (De Berardinis and others 1993, p.# 451). Local exhaust devices are not as effective as chemical fume hoods, so engineering and industrial hygiene professionals should be consulted to assist with selection or design for each specific application (NRC 1995, p.# 190).

Work Practices

Work practices are the most important element in controlling exposures.

Employees should understand the hazards associated with the procedures that they are performing, recognize the route through which they can be exposed to those hazards, select work practices that minimize exposures, and through training and experience acquire the discipline and skill necessary to sustain proficiency in the conduct of safe practices. Several categories of work practices should be considered:

- Practices to reduce the number of employees at risk of exposure.
  - Restrict access to the work area.
  - Provide warnings of hazards and advice about special requirements.
- Practices to reduce exposures by direct and indirect contact.
  - Keep hands away from mouth, nose, eyes, and skin.
  - Wash hands when contaminated and when work activity is completed.
  - Decontaminate work surfaces before and after work and after spills of a hazardous agent.
  - Use appropriate methods to decontaminate equipment, surfaces, and wastes.
  - Substitute less-hazardous materials for hazardous materials whenever possible.
  - Wear personal protection equipment (gloves, gowns, and eye protection).
- Practices to reduce percutaneous exposures.
  - Eliminate the use of sharp objects whenever possible.
  - Use needles with self-storing sheaths or those designed to protect the user.
  - Keep sharp objects in view and limit use to one open needle at a time.
  - Use appropriate gloves to prevent cuts and skin exposure.
  - Select products with puncture-resistant features whenever possible.
  - Use puncture-resistant containers for the disposal of sharps.
  - Handle animals with care and proper restraint to prevent scratches and bites.
- Practices to reduce exposure by ingestion.
  - Use automatic pipetting aids; never pipette by mouth.
  - Do not smoke, eat, or drink in work areas used for the care and use of research animals.
  - Keep hands and contaminated items away from mouth.
  - Protect mouth from splash and splatter hazards.
- Practices to reduce exposure by inhalation.
  - Use chemical fume hoods, biological safety cabinets, and other containment equipment to control inhalation hazards.
  - Handle fluids carefully to avoid spills and splashes and the generation of aerosols.
  - Use in-line HEPA filters to protect the vacuum system.

Handling and Transport of Animals

Safety precautions are needed during animal handling and animal transportation to prevent transmission of zoonotic agents to employees. Employees should wear personal protective equipment specifically chosen for the exposures that might be related to the animals being handled or transported. Safety concerns are relevant to all who have access to the animals being transported and those who receive and use them.

Personal Hygiene

Scrupulous attention to personal hygiene is essential for all personnel who care for and use research animals. They should wash their hands before and after handling animals and whenever protective gloves are removed. There should be no eating, drinking, smoking, application of cosmetics, or other activities that can increase the risk of ingesting hazardous materials or contaminating mucous membranes in animal care and animal use areas.

Housekeeping

All animal care areas, including areas in which hazardous materials are used or stored, should be kept neat and clean. Clutter can become contaminated and add to problems of employee exposure, area decontamination, and waste disposal. Work surfaces should be wiped with disinfectant before work begins, immediately after any spill, and at the end of the workday. Floors should be disinfected or decontaminated daily or weekly as appropriate to the potential hazards. Appropriate dust suppression methods should be routinely used. Wet mopping and the use of a HEPA-filtered vacuum cleaner are appropriate for suppressing dust.

Waste Disposal

Wastes need to be removed at scheduled intervals based on the amount of waste generated and the risk posed by the hazardous agents in the waste material. Planning is required to ensure that sufficient space is available for on-site collection, storage, treatment, and disposal of waste. The disposal of hazardous wastes is subject to federal, state, and local regulations. The environmental health and safety staff should stay informed of regulations, which change often. They should keep all on-site generators of hazardous waste informed of disposal procedures to ensure that they are in compliance with current requirements.

Restraint of Animals

Species specific safe techniques should be used to restrain animals (NRC 1996, p.# 11). Physical restraint might require more than one animal handler. Hand catching of nonhuman primates should be discouraged; use of a pole and capture collar is a safe alternative. The use of mechanical restraint devices or chemical restraints can reduce the potential for escape or injury when animals are being examined or handled. Employees should be aware that physical restraint can increase the inherent risks associated with the animal by intensifying excretions, secretions, and aggressive behavior of the animal.

Cleaning Cages

Caution should be used in removing animals from their cages before cage cleaning to avoid escape. Contaminated shavings, feces, urine, and other potentially biohazardous, contaminated, or allergenic materials should be removed with methods that protect the workers (NRC 1996, p.# 43-4). Biological safety cabinets have been designed as bedding dump stations to protect workers from hazardous aerosols that might be generated during cage cleaning. Protective clothing is required to protect workers from contact and percutaneous exposure. The eyes, face, and body should be protected during use of hazardous chemicals. Automatic cage washers pose several problems that should be addressed, including excess noise that might require hearing protection and ergonomic deficiencies that might contribute to back injuries and repetitive-motion injuries. Sharp edges on cages and ancillary equipment should be identified and eliminated. Heat in cage washing areas might require changes in ventilation and work practices to avoid excessive heat exposure. Employees should wear appropriate footwear and remain vigilant to the ever-present hazard of wet, slippery surfaces.

Personal Protective Equipment

The use of personal protective equipment is the final measure for controlling exposures to potentially hazardous agents. Personal protective equipment provides a physical barrier to hazardous materials that might otherwise come into contact with employees' skin, eyes, mucous membranes, and clothing. The equipment should protect the part of the body that is reasonably expected to come into contact with hazardous agents. Selection should be based on specific knowledge of the potential hazards, experience, and sound professional judgment.

Gloves are the most commonly used personal protective clothing. Latex, vinyl, or other appropriate protective gloves should be worn for handling potentially contaminated animals or hazardous materials. Care should be taken to ensure that the glove material provides an adequate barrier against the expected hazard. For example, nitrile or rubber gloves might be required to protect against some solvents, whereas thick leather would provide better protection against animal bites or scratches. Gloves should be long enough to cover the area to be protected.

Disposable vinyl or latex examination or surgical gloves should not be reused. Heavy duty rubber gloves will hold up well in cleaning and disinfecting; these are of the type commonly used for washing cages.

Uniforms, gowns, or laboratory coats are often provided to prevent contamination of animal care personnel by animal urine and feces. Such garb should not be worn outside the work area (unless it is covered). Protective clothing should be selected so that it provides an adequate barrier against the type and extent of exposure expected. For example, cage washing personnel might wear heavy rubber aprons to protect themselves when using strong detergents and cleaning agents. Safety shoes might be advisable for employees engaged in moving cage carts and other heavy equipment. Similar protective clothing might be needed by those who clean and disinfect animal rooms. The need to decontaminate and dispose of protective equipment is an important consideration in its selection. Reprocessing contaminated laundry can be more expensive than providing disposable gowns.

Face protection is advised if the eyes, nose, or mouth might be exposed through splashes or splatters of potentially hazardous agents. Safety glasses should be considered minimal eye protection and worn to prevent injury from projectiles, minor splashes, or contact of contaminated hands with eyes. Goggles or face shields might be needed for tasks involving infectious or hazardous liquids if there is a potential for splashing and splattering. Goggles or face shields are especially important when disinfectants and cleaning agents are used under pressure. Surgical masks also provide some protection of the mouth from splashes.

Respiratory protection might be necessary to control occupational exposures to aerosols. Employees who require respiratory protection should be enrolled in a respiratory program that is in compliance with OSHA standards. The selection and use of proper respiratory protection equipment should be coordinated through the environmental health and safety staff.

EDUCATION AND TRAINING

Occupational health and safety objectives of an institution can be achieved only if employees know the hazards associated with their work activities; understand how the hazards are controlled through institutional policies, engineering controls, work practices, and personal protective equipment; and have sufficient skills to execute safe work practices proficiently. All that requires a multifaceted education and training effort that addresses the full range of health and safety issues related to the care and use of research animals. Approaches for providing an education and training effort depend on the size, resources, animal species, research activities, staff experience, and technical expertise of the institution. However, successful programs have three common attributes:

- The occupational health and safety goals of the institution and how they will be achieved, including precise guidance on regulatory-compliance strategies, are clearly communicated to all employees. This function is commonly carried out by the environmental health and safety staff through formal orientation, distribution of written guidelines, and periodic refresher training.
- Employees are fully apprised of all relevant hazards and control strategies pertaining to their general work assignments. Information provided to employees is developed through the interaction of several key people, including a veterinarian or other professional familiar with zoonotic risks presented by the research animals, a health and safety professional who has knowledge of occupational hazards common to animal care and use and relevant hazard control strategies, and scientists who can assess the health risks associated with planned experimentation or research protocols. This interaction will define the knowledge needed by employees to protect themselves from hazards associated with their work and point to needs for further training.
- Supervisors in the animal care and research groups are actively involved in ensuring that their employees have acquired the necessary skills and attitudes to work safely. If deficiencies are present, on-the-job training supervised by an experienced employee is provided until appropriate standards of proficiency are demonstrated.

The involvement of scientists in the development of content for health and safety training is particularly important. Often, employees who routinely care for research animals are not placed under the direct supervision of the scientist who is responsible for the research project, so there can be gaps in communication of information that is necessary to protect the animal care staff or that could correct misperceptions about the risk of the research project. Such gaps could also place research animals at unnecessary risk. For example, research animals might be susceptible to disease when exposed to an animal care employee who is an active carrier of an infectious agent. The health and safety assessments of the scientists need to be conveyed both to the animal care staff and to their supervisors. Their understanding of the research objectives and the attendant hazards will help them to create and maintain a safe work environment in which the animal care staff can be integral and knowledgeable participants in the research activity.

Training should be a continuing process. It is best accomplished by identifying specific employees in a laboratory or animal care group to serve as a source of information, guidance, and instruction for their colleagues. The designated employees should be kept well informed of institutional health and safety requirements, safe practices, and relevant research and animal care hazards, and this requires a structured effort whereby the institutional experts in health and safety, animal care, and research interact with and advise them on all aspects of the institution's occupational health and safety program. Employees who serve in this way should have the recognition, confidence, and genuine support of their supervisors to carry out their important role effectively. The additional health and safety responsibilities are not likely to distract them substantially from the normal daily duties that they were hired to perform.

That approach has been successful in many research laboratories. Often, a laboratory manager oversees a laboratory's safety program and assigns specific aspects of the program, such as waste management or radiation safety, to other technical staff. The designated employees can serve as mentors and on-the-job trainers for new employees and provide guidance to more experienced workers as the need arises. These duties can be rotated among the experienced staff every several months—a practice that can quickly result in a highly informed and skilled workforce. Periodic group meetings are also helpful; they can serve as a forum for refresher training, provide opportunity for open discussion of safety concerns, and be used to review progress in achieving institutional health and safety goals.

An effective education and training program requires resources, administrative recordkeeping, and a mechanism for monitoring its efficiency, in addition to the interactive and mentoring efforts of key employees who provide relevant health and safety information. Investment of resources will produce a considerable return. A well-informed staff with safe work habits will minimize occupational injuries and illnesses. That in turn will reduce costs related to labor time, insurance, health care, disability, and legal actions.

Recordkeeping is an essential aspect of an education and training program. No program can succeed without knowledge of who needs what training and when such training has been provided. Training records are also required to satisfy specific requirements of federal and state environmental health and safety regulations. The institution's official responsible for ensuring maintenance of training records, usually the head of the environmental health and safety office, should strive to establish a simple system that presents the least administrative burden to everyone. A computer-based system should facilitate such an approach.

A wide variety of mechanisms exist for evaluating the success of the education and training program. Among these are site inspections, personnel reviews, injury and illness records, regulatory-compliance citations, and periodic questionnaires. The approach should be carefully designed and applied to provide information useful for both institution officials and employees.

EQUIPMENT PERFORMANCE

The value of engineering controls in protecting the health and safety of employees depends on the performance and operational integrity of the protective equipment. The environmental health and safety office should include programs for certifying and monitoring equipment to ensure that it is capable of providing the necessary protection and maintaining adequate performance.

The American National Standards Institute (ANSI) has published consensus guidelines for laboratory ventilation systems (ANSI Standard Z9.5-1993), which include recommendations regarding chemical fume hood performance. The ANSI standards are excellent reference documents and provide relevant guidance for engineering control of hazards in the care and use of research animals. The following ANSI recommendations refer specifically to chemical fume hoods:

- A routine performance test should be conducted on every fume hood at least once a year or whenever a substantial change has been made in the operational characteristics of the system.
- Each hood should maintain an average face velocity of 80-120 ft/min with no face-velocity measurement more than 20% greater or less than the average.
- New and remodeled hoods should be equipped with a flow-measuring device.

Biological safety cabinets should be tested and certified after installation and whenever a stationary cabinet is moved and should be recertified at least once a year (CDC-NIH 1995). Performance certification criteria have been established by the National Sanitation Foundation (NSF 1992).

Ultraviolet (UV) radiation of 254-nanometer (254-nm) wavelength may be used to control airborne and surface microorganisms in various locations in an animal care and research facility. The biocidal capacity of UV bulbs decreases with time and is adversely affected by contamination with dust or chemical films. They should be cleaned once a week and replaced on a regular schedule or monitored at least once a year to verify adequate performance (Fleming and others 1995, p.# 233).

HEPA filtration units require periodic monitoring to ensure filtration efficiency (NSF 1992). Performance tests should be conducted at least once a year. Appropriate controls or decontamination should be used during replacement and certification because filters can become contaminated with potentially infectious agents, toxic chemicals, or radioisotopes during use.

Charcoal filtration is occasionally used to control the environmental release of toxic materials or radionuclides. Performance is difficult to certify, and performance testing should be specific to the hazard that is being controlled (Shapiro 1990, p.# 331). Performance should be monitored either by using continuous monitoring instruments, which are calibrated to the chemicals of concern and placed downstream of the exhaust-filter bed, or by periodically sampling the discharge air. An acceptable alternative to monitoring charcoal filtering systems is to replace the filters at established intervals that are based on their calculated effective life. Ductless fume hoods that use mainly activated charcoal filters should not be used for protection from volatile toxic compounds (Keimig and others 1991; NRC 1995, p.# 185).

Ventilation system performance should be checked periodically to document adequacy of room air exchanges and air pressure gradients in accordance with authoritative guidelines (NRC 1996). Air pressure gradients indicate airflow relationships; the frequency of monitoring them should be based on the degree of risk associated with the hazardous materials being used. Continuous readout monitoring instruments might be appropriate to provide instantaneous performance information in high containment facilities.

Effluent monitoring might be required by local ordinances designed to protect the sewage treatment works of the municipality. Specific, periodic monitoring might be required for ensuring compliance with discharge limits for chemical, biological, or radiological agents. Additional monitoring could also be necessary to demonstrate adequate control after accidental spills or releases of materials that might have entered the sewage system.

Validation and verification are important aspects of autoclave performance testing. The use of biological indicators that contain bacterial spores is an effective method of validating sterilization cycles for various load types. Monitoring of autoclave operational measures (temperature, pressure, and time) can verify performance routinely.

Fire protection systems and equipment (such as fire extinguishers) should be inspected and tested periodically to ensure operational integrity. Insurance companies and local fire authorities generally specify the frequency with which these inspections and tests should be performed.

INFORMATION MANAGEMENT

Rapid access to employee-specific exposure information is increasingly important for efficient safety management. Documentation of occupational exposures, safety training, medical surveillance, and work related injury and illness is important for evaluating the occupational health and safety program of the institution, promoting health and safety, identifying new occupational risks, ensuring the cost effectiveness of program activities, and achieving regulatory compliance. On-line access to relevant health and safety information could improve the management and performance of occupational health and safety programs. It would facilitate the exchange of information between environmental health and safety, occupational health, animal care and use, and research staffs. On-line interactions could make it practical to develop records that are specific for each research protocol and that contain information relevant to each potentially exposed employee. Table 6-1 lists examples of information elements that can be shared in an occupational health and safety information-management network. Confidentiality and limited access to some kinds of information should be ensured.

TABLE 6-1 Occupational Health and Safety Information-Management Network

Activity	Information Provided by Activity	Information Received by Activity
Animal care and use	Job profile Project risk data Training records	List of employees at risk Employment risk indicators Exposure and monitoring data Health evaluation data Health surveillance data Health surveillance schedules Material safety data sheets Risk assessment data Training schedules
Research	Job profile Project risk data Training records	List of employees at risk Employment risk indicators Exposure and monitoring data Health evaluation data Health surveillance data Health surveillance schedules Material safety data sheets Risk assessment data Training schedules
Environmental health and safety	Accident and injury investigation data Employment risk indicators Exposure and monitoring data Material safety data sheets Risk assessment data Training schedules	List of employees at risk Hazardous materials purchasing data Health evaluation data Health surveillance data Job classification and position descriptions Job profile OSHA 200 log data Project risk data Training records Worker compensation data
Occupational health	Health evaluation data Health surveillance information Health surveillance schedules	Accident and injury investigation data List of employees at risk Employment risk indicators Exposure and monitoring data Job profile Material safety data sheets OSHA 200 log data Risk assessment data Worker compensation data

continued on next page

TABLE 6-1 Continued

Activity	Information Provided by Activity	Information Received by Activity
Administration and management	List of employees at risk Hazardous materials purchasing data Job classification and position descriptions OSHA 200 log data Worker compensation data	Accident and injury investigation data Employment risk indicators Job profile Project risk data

Computer links with other institutions through external networks, such as electronic mail, are useful for obtaining current health and safety information. Numerous safety bulletin boards are available for communicating with health and safety personnel throughout the world. Other specialty boards provide easy access to information regarding infectious, toxic, and radiological hazards.

EMERGENCY PROCEDURES

All institutions should have emergency response plans. Emergency situations will occur, and they require a rapid, coordinated response to minimize harm to personnel and facilities. A rapid and appropriate response is not possible without an institutionally adopted written and tested plan.

An emergency response plan provides a structure for effective response by defining employee responsibilities, interactions between responding personnel, the sequence of response procedures, and availability of emergency equipment. The complexity of the plan will be dictated by the diversity of emergencies that are considered possible and the institutional capacity and ability to respond to emergencies with on-site personnel. Environmental health and safety personnel should be readily available to coordinate response efforts, and all off-site emergency responders should be well educated in the unique hazards and situations that might occur. All on-site employees should know their roles in responding to emergency situations.

The planning process should follow a logical progression that begins with identification of the types of emergency situations that are most likely to occur. That information is used to determine who should respond to each identified situation. Equipment requirements need to be determined by those who will respond to emergencies. After adoption of the written plan by the institution, training will need to be conducted and drills performed to test its efficacy.

The emergency response team for an animal facility should either include or have rapid access to health and safety, veterinary, and animal care personnel. Hazards related to the animal care and use program should be known to ensure that adequate equipment and training are available. All personnel involved in emergency response should know the limitations of their training and equipment and not perform activities for which they have not been trained.

Typically, the hierarchy for response will be to protect personnel, then animals, and finally the animal care facility and surrounding buildings. The plan should include provisions for moving or relocating animals to temporary housing facilities. The temporary facilities should be adequately equipped to address the needs of the different species that might require relocation.

Medical personnel should receive specific information on the unique hazards related to emergency response procedures in the animal facility. They will need to be prepared to support the potential exposures and injuries related to emergency responses. All injuries managed by medical personnel should be included in the system for reporting work related injuries and illnesses.

PROGRAM EVALUATION

The quality and effectiveness of an institution's occupational health and safety program can be sustained only through periodic evaluations of the program and a commitment to respond to changing circumstances. Evaluations should be performed at the request of the senior official of the institution, who should act on the findings. The source of the request establishes institutionwide interest in the evaluations, involves top administration in the deliberations, and ensures close communication between the evaluation group and institutional officers.

Members of the evaluation group should be appointed by the senior official of the institution. The group should include appointees from each of the major activities in the occupational health and safety program. Individually, the members should be recognized by their peers as persons of good judgment. All should have a personal commitment to the objectives of the occupational health and safety program. Chairpersons of relevant committees should participate in the evaluation, and the managers of the environmental health and safety and occupational health activities should serve as resources for the group.

The evaluation should be based on objective data that will help in measuring the effectiveness of the program in reducing occupational risks to an acceptable minimum. Three general subjects should be emphasized: the institution's injury and illness experience, its regulatory-compliance performance, and the results of efforts to promote health and safety through continuing interactions among the major participants in the occupational health and safety program.

The data sources should include the results of exposure monitoring if performed for any purpose, worker compensation records, OSHA recordable injuries and illnesses, results of special health and safety studies or investigations, training records, minutes and reports of institutional health and safety committees and any related actions taken by the IACUC, and results of inspections conducted by regulatory agencies. Some institutions perform self-audits to identify deficiencies or recommend improvements in their environmental, health, and safety regulatory-compliance activities; this approach is viewed favorably by most regulatory agencies.

The best measure of the effectiveness of interactions among the major participants in the program is whether the health and safety policies, rules, and recommended practices are relevant to the hazards that are present and can be implemented in a practical manner. That might well be a subjective assessment, but it is exceedingly important. Relevance and practicability influence attitudes, and positive attitudes toward the occupational health and safety program minimize risks.

7

Occupational Health-Care Services

Part of the mission of an occupational health and safety program is to foster the prevention of occupationally acquired illnesses and injuries, the early recognition of health alterations due to occupational exposures, and the treatment and management of occupationally acquired illnesses and injuries. Ideally, the occupational health element of the occupational health and safety program encompasses a multifaceted occupational health-care service that complements the overall program and meets the institution's need for productivity and cost effectiveness. The occupational health-care service works within the occupational health and safety program to ensure that risks associated with the use of research animals are kept to an acceptable minimum.

The wide variety of acceptable arrangements for providing occupational health-care services reflects the variation in institutional needs and resources, including the size of the animal care and use program, the nature of the risks, and access to occupational health-care services. The services are provided by groups or individuals that have training or experience in occupational health. The providers include physicians, nurse practitioners, physician's assistants, and nurses. Each institution should select or contract for appropriate professional guidance and occupational health-care services to meet the occupational-health needs of its employees.

This chapter focuses on occupational health-care services appropriate for employees engaged in the care and use of research animals. The term *employee* is used as a functional term and refers to all persons whose duties place them near research animals, their derived products, and their tissues. The term is intended to include animal-care personnel, investigators and their technical staff, students and other trainees, volunteers, engineers, housekeepers, security officers, and maintenance personnel as appropriate. The occupational health-care services needed for employees vary with the health risk associated with their animal-related research or support activities. Institutions should strive for consistency in the occupational health-care services provided for employees at comparable risk.

Institutions often do not provide occupational health-care services for contract employees who participate in an animal care and use program. If an institution does not provide services, it should confirm that contractors understand and accept their responsibility for the health and safety of contract workers. The contractor should provide occupational health-care services to its employees that are consistent with those provided by the institution to its employees. The institution with responsibility for the space where the contract workers work is obliged to communicate to the contractor the hazards and risks present in the worksite and rules and procedures for the maintenance of a safe environment, and the contractor is obliged to follow and enforce safe work practices.

FEDERAL REQUIREMENTS AND GUIDELINES FOR OCCUPATIONAL HEALTH-CARE SERVICES

The Occupational Safety and Health Act mandates that employers provide a safe and healthful workplace for their employees. Occupational health-care services might be required for the institution to meet its responsibilities under the general-duty clause of the act and those specified in health standards promulgated by the Occupational Safety and Health Administration (OSHA). For example, an institution would be required under the OSHA bloodborne-pathogens standard (29 CFR 1910.1030) to provide hepatitis B vaccinations to employees who handle blood, organs, or other tissues from experimental animals infected with HBV and to make available to an employee a confidential medical evaluation immediately after exposure to animal tissues that are contaminated with a bloodborne pathogen. The OSHA standard on occupational exposure to hazardous chemicals in laboratories (29 CFR 1910.1450) requires medical surveillance when exposure-monitoring reveals an exposure routinely above the action level for an OSHA-regulated substance, e.g., a time-weighted average (TWA) of 0.75 ppm or a short-term exposure level (STEL) of 2.0 ppm for formaldehyde (29 CFR 1910.1048). However, OSHA action levels are unlikely to be exceeded in an animal care and use setting.

The *Public Health Service Policy on Humane Care and Use of Laboratory Animals*, which was revised in response to the Health Research Extension Act of 1986, requires institutions that receive federal funds to provide occupational health-care services to employees who work in laboratory animal facilities and have substantial animal contact. The National Institutes of Health *Guidelines for Research Involving Recombinant DNA Molecules* (NIH 1994) requires institutions that receive NIH support for recombinant-DNA research to provide occupational health-care services to employees engaged in animal research involving viable recombinant-DNA-containing microorganisms that need biosafety level 3 or greater containment. The NIH guidelines suggest that, for this level of risk, occupational health-care services would include records of agents handled, active investigation of relevant illnesses, and maintenance of serial serum samples for monitoring serological changes that might result from employees' work experience.

Specific occupational health-care services are recommended in *Biosafety in Microbiological and Biomedical Laboratories* (CDC-NIH 1993) for employees engaged in research programs that involve experimentally or naturally infected vertebrate animals. These services are summarized in Table 7-1. This authoritative source should be consulted for further detail and guidance regarding the application of these recommendations to specific research situations.

ASSESSMENT OF HEALTH RISKS

Employees who are involved in the care and use of research animals might face health risks for which specific health-care services should be provided. In most cases, effective use of good animal-care and occupational health and safety practices will be sufficient to protect the health of employees. But some work activities create higher risks of occupational injury and illness (such as handling of heavy cages, e.g., back injuries; direct handling of macaques, e.g., B-virus exposure; and removing litter from cages, e.g., increased exposure to allergens). Institutions are obliged to determine which activities and positions place employees at higher risks and to provide the necessary health-care services for them.

Substantial contact with research animals or their tissues is an important consideration in assessing health risks, although it is inadequate as the sole criterion for assessing risk. Several aspects of a job merit consideration, including exposure intensity, exposure frequency, the hazards associated with the animal being handled, the hazardous properties of the agents that are used in research, the susceptibility of the individual, and the occupational-health history of previous employees. Ultimately, the determination of risk and of the need for health-care services is a matter of professional judgment, especially when the frequency and intensity of exposures to hazards are borderline. Risk associated with the care and use of research animals can be assessed in accordance with the criteria, classifications, and kinds of information listed in Table 7-2.

RESPONSIBILITIES OF AN OCCUPATIONAL HEALTH-CARE SERVICE

An occupational health-care service has various important responsibilities:

- To commit to developing a detailed knowledge of the occupational haz-

TABLE 7-1 Federal Recommendations for Occupational Health-Care Services for Research Programs That Involve Experimentally or Naturally Infected Vertebrate Animals

Practice	Recommendation
Limiting access	Employees who are highly susceptible to infection with the agent under study or for whom infection might be unusually hazardous should not work in areas where the agent is handled or where vertebrate animals that are experimentally or naturally infected with the agent are used and cared for.
Collection and storage of baseline serum samples	Biosafety level 2: When appropriate, considering the agents handled (e.g., where there is substantial risk of occupationally acquired infection with the agent under study and methods are available to measure immunologic response to the agent). Biosafety level 3: For all employees who have access to areas where the agent under study is handled or where vertebrate animals that are experimentally or naturally infected with the agent are used and cared for and where methods are available to measure immunologic response to the agent.
Serological surveillance	Periodic collection and testing of serum samples for at-risk employees is recommended where there is substantial risk of occupationally acquired infection with the agent under study and methods are available to measure immunologic response to the agent. Testing of the sample should be conducted at each collection and the results communicated to the employee.
Immunization	Immunizations are recommended for clearly identified at-risk employees where a safe and effective vaccine or toxoid exists (e.g., vaccines against hepatitis B, yellow fever, rabies, and poliomyelitis, and toxoids against diphtheria and tetanus). Decisions for giving less-efficacious vaccines, those associated with high rates of local or systemic reactions, or those which produce increasingly severe reactions with repeated use should be carefully considered.
Screening tests	Skin testing with purified protein derivative (PPD) of previously skin-test-negative at-risk employees is recommended.

Source: CDC-NIH 1993.

TABLE 7-2 Assessment of Risk Associated with Animal-Related Research

Criterion	Possible Classifications	Information Sources
Exposure intensity	High Medium Low Absent	Job profile, environmental health and safety assessment, employee history
Exposure frequency	8 h/wk or more Less than 8 h/wk No direct contact Never	Job profile, environmental health and safety assessment, employee history
Hazards posed by animals	Severe illness Moderate illness Mild illness Illness unlikely	Institutional veterinarian
Hazards posed by materials used in or with animals	Severe illness Moderate illness Mild illness Illness unlikely	Material-safety data sheets; CDC-NIH agent summary statements; radiation-, chemical-, and biological-safety committees; environmental health and safety staff
Susceptibility of employee	Direct threat[a] Permanent increase Temporary increase	Medical evaluation, review of personal medical records
Expected incidence or prevalence	High Medium Low Absent	Published reports, industry experience
History of occupational illness or injury in the position or workplace	Severe Moderate Mild None	Worker-compensation reports, OSHA 200 log
Regulatory requirements	Required for any contact Professional judgment permitted	Environmental health and safety office, consultants, risk managers

[a]Reasonable probability of substantial harm. Americans with Disabilities Act of 1990 (PL 101-336).

ards of employees and an understanding of the temporal and spatial distribution of those hazards. These are referred to below as "that knowledge" and "that understanding."

- To understand the medical presentation of occupational illness and injuries for which employees are at risk.
- To understand the characteristics of the workforce, the nature of sensitivity or susceptibility factors among members of the workforce, and how these factors affect the ability of employees to perform their tasks.
- To apply that knowledge to an understanding of how employment presents a direct threat to employees' health.
- To communicate that understanding to the health, safety, and management teams to assist them in making program decisions that are based on the best available medical knowledge.
- To communicate that understanding to potential and current employees so that they can decide whether to accept potential hazards.
- To communicate the necessary medical information in the event of an occupational illness or injury in a timely fashion to persons with a need to know, including human-resources, worker-compensation, health and safety, and supervisory personnel.
- To strive to maintain objectivity in the face of conflicts that occur because of the occurrence of work-related illness or injury.
- To educate employees about early warning signs of occupational illness or injury that should prompt medical action or evaluation.
- To provide the institution a considered judgment, based whenever possible on aggregate information, as to the status of occupation-related illness and injury among employees.
- To participate in the identification of employees at high risk because of animal-related research.

The effectiveness with which those responsibilities are carried out depends on the health-care provider's knowledge of the employee health risks associated with the care and use of research animals at the institution. It also depends on the opportunity to foster genuine collaboration among all program activities of the institution that manage, support, and conduct the animal care and use program.

There is a major need for a basic, accessible body of knowledge about health risks to employees. Health-care providers need to have appropriate training and experience to establish and maintain an effective health-care service as part of an occupational health and safety program for employees involved in the care and use of research animals. The information needed to conduct an occupational-health program is typically acquired from many sources. It is essential that veterinarians, investigators, and environmental health and safety professionals participate in the orientation and continuing training of health-care providers about zoonoses, exposures, illnesses, and other health risks associated with the care and use of research animals. Infectious-disease specialists, allergists, dermatologists, and pulmonologists might also have to be consulted about aspects of employee health.

ACTIVITIES OF AN OCCUPATIONAL HEALTH-CARE SERVICE

The selection of occupational health-care services is based on knowledge of occupational hazards, the nature of health risks associated with animal care and use activities at the institution, and the diversity of employees, the work environment, and the mission of the institution. An occupational health-care service that provides comprehensive health-care services to all employees engaged in the care and use of research animals without consideration of employee risks is expensive and might not convey the understanding that employees must have to minimize occupational-health risks. Greater value comes from occupational health-care services that are selectively and judiciously based on work activities that place employees at risk of occupational injury or illness. For example, a preplacement medical evaluation usually consists of a review of functional demands of a position, hazards associated with the animal species involved, potential experimental hazards, and the employee's medical history. Such an evaluation makes good sense if an employee is being assigned duties that require heavy lifting, the handling of animals that are known to be potential sources of zoonotic infections, the cleaning of cages, or the handling of bloodborne pathogens. But it would not be prudent or cost-effective to perform preplacement evaluations of employees only on the basis of substantial contact with research animals, because resources would be directed where hazards do not exist.

The occupational health-care services can include preplacement medical evaluations, periodic health evaluations, episodic health evaluations, analyses of adverse health outcomes, medical management of worker-compensation cases, immunizations, medical recordkeeping, serum-banking, exit evaluations, and nonoccupational health care. The value and relevance of those activities for employees at risk are discussed in the sections that follow. No activity should be selected for inclusion in an institution's occupational health-care service without consultation with environmental health and safety professionals and discussion with representatives of the research and animal care and use programs.

Identification of Persons at High Risk

The institution should identify employees at risk because of animal-related research and determine who should participate in the various activities provided by the health-care service. Categories of employees whose activities should be reviewed are investigators, technicians, animal-facility operators, clerical and other support personnel, students, trainees, visitors, maintenance and housekeeping personnel, engineers, and facility-management technicians. The service components that are needed vary with the nature and intensity of the risk.

Interaction with Environmental Health and Safety Staff

Interaction between occupational health-care service staff and environmental health and safety staff is necessary to develop workplace-exposure information needed for health-care services. Such interaction constitutes a process for alerting environmental health and safety professionals to hazards that might require additional control. This interaction is also important for assessing risks associated with activities related to animal research and helps to establish criteria for selecting employees who will routinely receive health-care services.

Preplacement (Preassignment) Medical Evaluations

The preplacement evaluation serves several functions in the occupational health-care service. Every employee who is identified to participate in various activities of the health-care service and is subject to substantial risk in the animal care and use program should undergo a preplacement medical evaluation. It establishes baseline health information on employees before exposure to the risks associated with animal-related research. Pre-existing conditions that can affect an employee's capability to perform the essential functions of his or her position without risk of substantial harm might be identified. Another function is to discuss medical conditions that might alter an employee's exposure-risk profile; these could include current conditions (such as tuberculosis) and possible future conditions (such as pregnancy in women of child-bearing age). Medical conditions that could temporarily alter fitness for duty or require on-site emergency treatment (such as diabetic hypoglycemia and epileptic seizures) can be noted, and appropriate contingency plans can be made. The preplacement medical evaluation also presents an opportunity for education about potential hazards in the workplace, the need for accommodation or personal protection, and medical symptoms that should prompt an employee to seek occupational-health evaluation between routine visits.

Periodic Health Evaluations

Scheduled, periodic health evaluations are often a key component of occupational-health programs. They are most useful when carefully designed to obtain information that can be used to verify the success of the occupational health and safety program in reducing occupational illness and injury. The components and frequency of evaluations depend on the nature of potential hazards. Symptoms of health alterations that are of insufficient severity to be labeled disease can prompt preventive measures. Knowledgeable and experienced people—including representatives of worker compensation, environmental health and safety, personnel (human resources), and the occupational health-care service—should determine the need for and design of periodic health evaluations. There should be a schedule for the re-evaluation of previous decisions, the interval for which depends on changes in exposure or workforce characteristics, injury and illness experience, and the availability of new guidance regarding good occupational-health practice.

Physical examinations need not be a routine part of periodic medical evaluations. Periodic workplace physical examinations are typically performed on healthy persons and rarely alter judgments about their fitness for duty. Resources can be better spent on aggregating and analyzing health-status information, performing worksite tours, and tailoring health programs to be specific to the circumstances of each worksite. The time spent with an employee in a medical evaluation might be better spent in taking a careful history based on a knowledge of worksite risks, informing the employees of the nature of hazards and the means of protecting against them, and warning signals of illness.

Episodic Health Evaluations

Persistent symptoms, symptoms that indicate the onset of a work-related illness, or the occurrence of a work-related injury should prompt appropriate medical evaluation and care. A physical examination focused on the chief complaint is typically needed as a routine part of an episodic health evaluation. The results of some evaluations (such as the finding of an eye injury) are referred directly to specialists, and a mechanism is needed to make the health-care service staff and the environmental health and safety staff also aware of them. As a general rule, any event that leads to medical evaluation and any loss of work time that is thought to be work-related should be reported to the occupational-health information system (BLS 1986).

Recognition, Evaluation, Recording, and Followup of Adverse Health Outcomes

The incidence and prevalence of medical symptoms, injuries, or illnesses should be assessed periodically. Several mechanisms are used to recognize adverse health risks and adverse health outcomes. Incident reports are completed when medical symptoms occur as the result of a workplace event or exposure. They should be reviewed by the health-care service to determine whether medical evaluation is needed; the information should also be reviewed by the environmental health and safety staff to determine whether their involvement is needed.

"Near-miss reports" may be prepared by employees when equipment malfunction or performance error almost results in an accident or substantial exposure. Near-miss reports are usually kept by the environmental health and safety staff but can be reviewed by health-care providers.

Medical Management of Worker-Compensation Cases

The management and treatment of worker-compensation cases by the occupational health-care service might be an effective way to reduce incidence, severity, and costs of occupational injuries and illnesses (McGrail and others 1995). This service can provide closer monitoring of an employee's ability to return to work than an outside provider unfamiliar with the work setting. Return-to-work examinations allow for review of injuries and illnesses (work-related or personal) not being followed by the occupational health-care service and can facilitate an appropriate and safe return to the worksite.

Immunization

Immunization programs are an accepted method of protecting people from some infectious diseases. The decision to immunize an employee should be made because of a clearly defined, recognized risk at the time of preplacement, periodic, or episodic health evaluations (guidance for administration of specific vaccines and toxoids—such as for hepatitis B, rabies, and tetanus—is provided by the Public Health Service Advisory Committee on Immunization Practices (IPAC 1996).

Medical Recordkeeping

It is the responsibility of the employer to maintain medical records related to an employee's participation in a health-care service activity. Many employers delegate that responsibility to a contract medical service, but there should be a provision for transfer of records if the contractual arrangement terminates.

Aggregation of occupational-health data is commonly overlooked. Preplacement and periodic health evaluations are performed on many workers, but their results are seldom analyzed in the aggregate for informational purposes. The information derived from aggregate data can be of great use in guiding program decisions. Consultation with epidemiologists can be useful because they understand how and why information should be aggregated.

Serum-Banking

Serum-banking is the collection and frozen storage of serum samples drawn from employees who might be at risk for occupationally acquired infection. Typically, the purpose of the program is to give the institution the ability to compare serum obtained after an acute illness or exposure with serum obtained before the illness or exposure began. Although serum-banking has generally been regarded as a standard component of occupational-health programs, it should be conducted only when there is a clear reason for obtaining the specimens and there is a plan to analyze the data as part of a risk-assessment strategy. CDC and NIH (CDC-NIH 1993) recommend serum-banking and serologic surveillance when a substantial risk of occupational illness is associated with an agent under study and methods are available to measure immunologic response to the agent (see Table 7-1).

Substantial issues should be considered in advance of instituting a serum-banking program, including chain of custody, confidentiality, identification and handling of samples, retention, potential deterioration of sample quality over time, and cost. The program should include informed consent of employees and allow them to decline to participate. The collection and storage of employee serum should not be performed in the absence of a functioning occupational health and safety program.

Exit Evaluations

An exit evaluation is defined as a medical evaluation performed when an employee terminates employment. Its purpose is to determine the employee's health status when exposure to potential hazards ceases. Such an evaluation has potential value for medical and legal reasons. As a practical matter, however, few employees are interested in undergoing evaluations when they leave an employer; after the final paycheck has been disbursed, there are few incentives for the employee to return. It is unlikely that information useful to an occupational health and safety program will be obtained from exit interviews.

Nonoccupational Health Care

The occupational health-care service should not be the source of primary medical care for employees. Its use as such a source is discouraged because it diverts resources from aspects of the program aimed at reducing workplace health risks.

Some employers choose to use the occupational health-care service for general health promotion, such as blood-pressure measurement, cholesterol screening, and education about healthy lifestyles. The enthusiasm for that kind of promotion should be tempered by an honest assessment of the institution's resources that are available for occupational health.

PROGRAM EVALUATION

Evaluation of the adequacy of a health-care service should focus on whether the health-care providers meet legal requirements and ethical guidelines, accomplish the mission of the occupational-health program, recognize the essential elements of the health-care service, and deliver the appropriate components of the service.

The following conditions are indicators that a program is adequate:

- Health-care providers tour the facility and are knowledgeable about the workplace-hazard profile.
- The health-care service is aware of the occupational-health profiles of employees as reflected in the worker-compensation claims experience, the OSHA 200 log, first-aid reports, and incident reports.
- Medical histories elicit risk-related events (such as the frequency and severity of animal bites).
- The health-care service requests consultation from the environmental health and safety staff in the case of health alterations or occupational disease or injury.
- The health-care service participates in the development of activities of the occupational health and safety program.
- The health-care service provides information to the institution about the occurrence of work-related illnesses and injuries.

References

ACGIH (American Conference of Governmental Industrial Hygienists) 1994. P. 100 in Threshold Limit Values (TLVs) for Chemical Substances and Physical Agents and Biological Exposure Indices. American Conference of Governmental Industrial Hygienists.

Agrup, G., L. Belin, L. Sjostedt, and S. Skerfving. 1986. Allergy to laboratory animals in laboratory technicians and animal keepers. Br. J. Ind. Med. 43:192-98.

Alexander, A. D. 1984. Leptospirosis in laboratory mice. Science 224:1158.

Alvarez-Cuesta, E., J. Cuesta-Herranz, J. Puyana-Ruiz, C. Cuesta-Herranz, and A. Blanco-Quirós. 1994. Monoclonal antibody-standardized cat extract immunotherapy: risk-benefit effects from a double-blind placebo study. J. Allergy Clin. Immunol. 93:556-66.

American Thoracic Society. 1990. Diagnostic standards and classification of tuberculosis. Am. Rev. Respir. Dis. 142:725-35.

Anderson, B. C. 1982. Cryptosporidiosis: a review. J. Am. Vet. Med. Assoc. 180:1455-7.

Anderson, L. C., S. L. Leary, and P. J. Manning. 1983. Rat-bite fever in animal research laboratory personnel. Lab. Anim. Sci. 33:292-4.

Anderson, M. C., H. Baer, and J. L. Ohman, Jr. 1985. A comparative study of the allergies of cat, urine, serum, saliva, and pelt. J. Allergy Clin. Immunol. 76:563-9.

Andrei, G., and E. De Clercq. 1993. Molecular approaches for the treatment of hemorrhagic fever virus infections. Antiviral Res. 22:45-75.

ANSI (American National Standards Institute). 1986. Standard Z-136.1-1986. Safe Use of Lasers. New York: American National Standards Institute

ASM (American Society of Microbiology). 1995. Laboratory Safety: Principles and Practices. Washington, D.C.:ASM Press.

Baldo, B. A. 1993. Allergenicity of the cat flea. Clin. Exp. Allergy 23:347-9.

Barkin, R. M., J. C. Guckian, and J. W. Glosser. 1974. Infection by *Leptospirum ballum*: a laboratory associated case. South. Med. J. 67:155-76.

Barkley, W. E., and J. H. Richardson. 1984. The control of biohazards associated with the use of experimental animals. Pp. 595-602 in Laboratory Animal Medicine, J. G. Fox, B. J. Cohen, and F. M. Loew, eds. Orlando: Academic Press.

Benenson, A. S., ed. 1995a. Ebola-Marburg virus diseases. Pp. 159-60 in Control of Communicable Diseases Manual. 16th edition. Washington, D.C.: American Public Health Association.

Benenson, A. S., ed. 1995b. Other infections associated with animal bites. Pp. 84-5 in Control of Communicable Diseases Manual. 16th edition. Washington, D.C.: American Public Health Association.

Benenson, A. S., ed. 1995c. Giardiasis. Pp. 202-204 in Control of Communicable Diseases Manual. 16th edition. Washington, D.C.: American Public Health Association.

Bennett, B. T., C. R. Abee, and R. Hendrickson, eds. 1995. Nonhuman Primates in Biomedical Research: Biology and Management. San Diego, Calif.: Academic Press.

Benson, P. M., S. L. Malane, R. Banks, C. B. Hicks, and J. Hilliard. 1989. B virus (*Herpesvirus simiae*) and human infection. Arch. Dermatol. 125:1247-8.

Bernard, K. W., G. L. Parham, W. G. Winkler, and C. G. Helmick. 1982. Q fever control measures: recommendations for research facilities using sheep. Inf. Control. 3:461-5.

Bhatt, P. N., R. O. Jacoby, and S. W. Barthold. 1986. Contamination of transplantable murine tumors with lymphocytic choriomeningitis virus. Lab. Anim. Sci. 36:136-9.

Biery, T. L. 1977. Venomous Arthropod Handbook. Washington, D.C.: US Government Printing Office.

Bland, S. M., R. Evans, III, and J. C. Rivera. 1987. Allergy to laboratory animals in health care personnel. Occup. Med. 2:525-46.

Blaser, M. J., F. M. LaForce, and N. A. Wilson. 1980. Reservoirs of human campylobacterosis. J. Infect. Dis. 141:665.

Blaser, M. J. 1990. Campylobacter species. Pp. 1649-1656 in Principles and Practices of Infectious Diesase, G. L. Mandell, D. R. Gordon and J. E. Bennett, eds. New York: Churchill Livingstone.

Blumenthal, J. B., and M. N. Blumenthal. 1996. Immunogenetics of Allergy and Asthma. Immunol. Allergy Clin. No. Amer. 16:517-34.

Bothan, P. A., G. E. Davies, and E. L. Teasdale. 1987. Allergy to laboratory animals: a prospective study of its incidence and of the influence of atopy on its development. Br. J. Ind. Med. 44:627-32.

Bowen, G. A., C. H. Calisher, W. G. Winkler, A. L. Kraus, E. H. Fowler, R. H. Garman, D. W. Fraser, and A. R. Hinman. 1975. Laboratory studies of a lymphocytic choriomeningitis virus outbreak in man and laboratory animals. Am. J. Epidemiol. 102:233-40.

Bryant, E. 1984. Biology and diseases of birds. Pp. 400-26 in Laboratory Animal Medicine, J. G. Fox, B. J. Cohen, and F. M. Loew, eds. Orlando: Academic Press.

BLS (Bureau of Labor Statistics) U.S. Department of Labor. 1986. Recordkeeping Guidelines for Occupational Injuries and Illness. Washington, D.C.: US Government Printing Office.

Butler, T. 1990. Yersinia species (including plague). Pp. 1748-56 in Principles and Practices of Infectious Diseases, G. L. Mandell, D. R. Gordon, and J. E. Bennett, eds. New York: Churchill Livingstone.

CCAC (Canadian Council on Animal Care) 1993. Approaches to the Design and Development of Cost-Effective Laboratory Facilities. CCAC proceedings. Ottawa, Ontario, Canada: CCAC. 273 pp.

CDC (Centers for Disease Control). 1976. Unpublished Data. Center for Infectious Diseases. US Department of Health, Education, and Welfare, Public Health Service.

CDC (Centers for Disease Control). 1979. Q fever at a university research center-California. MMWR 20(13):147-8.

CDC (Centers for Disease Control). 1987. B-virus infection in humans-Florida. MMWR 36:289-96.

CDC (Centers for Disease Control). 1988. Guidelines to prevent simian immunodeficiency virus infection in laboratory workers and animal handlers. MMWR 37:693-4, 699-704.

CDC (Centers for Disease Control). 1989a. B-virus infection in humans-Michigan. MMWR 38:453-4.

CDC (Centers for Disease Control). 1989b. Ebola virus infection in imported primates-Virginia. MMWR 38:831-2.

CDC (Centers for Disease Control). 1990. Update: Ebola-related filovirus infection in nonhuman primates and interim guidelines for handling nonhuman primates during transit and quarantine. MMWR 39:22-30.

CDC (Centers for Disease Control). 1992a. Seroconversion to simian immunodeficiency virus in two laboratory workers. MMWR 41:678-81.

CDC (Centers for Disease Control). 1992b. Anonymous survey for simian immunodeficiency virus (SIV) seropositivity in SIV-laboratory researchers-United States, 1992. MMWR 41:814-5.

CDC (Centers for Disease Control and Prevention). 1993a. Hantavirus infection-southwestern United States: Interim recommendations for risk reduction. MMWR 42:(No. RR-11).

CDC (Centers for Disease Control and Prevention). 1993b. Update: outbreak of hantavirus infection-southwestern United States. MMWR 42:441-3.

CDC (Centers for Disease Control and Prevention). 1993c. Tuberculosis in imported nonhuman primates-United States, June 1990-May 1993. MMWR 42:572-5.

CDC (Centers for Disease Control and Prevention). 1994a. Guidelines for preventing the transmission of Mycobacterium tuberculosis in health care facilities. MMWR 43:(No. RR-13).

CDC (Centers for Disease Control and Prevention). 1994b. Laboratory Management of Agents Associated with Hantavirus Pulmonary Syndrome: Interium Biosafety Guidelines. MMWR 43:(No RR-7).

CDC (Centers for Disease Control and Prevention). 1995. Methods for Trapping and Sampling Small Mammals for Virologic Testing. Atlanta: CDC.

CDC (Centers for Disease Control and Prevention). 1996. Guidelines for Working with Rodents Potentially Infected with Hantavirus. Atlanta: CDC.

CDC-NIH (Centers for Disease Control and Prevention-National Institutes of Health). 1993. Biosafety in Microbiological and Biomedical Laboratories, 3rd edition. HHS Publication No. (CDC) 93-8395, Washington, D.C.: US Government Printing Office.

CDC-NIH (Centers for Disease Control and Prevention-National Institutes of Health). 1995. Primary Containment for Biohazards: Selection, Installation and Use of Biological Safety Cabinets. Washington, D.C.: US Government Printing Office.

CFR (Code of Federal Regulations). Title 29 Part 72. Centers for Disease Control. Interstate Shipment of Etiologic Agents FR 45 (141) Monday, July 21, 1980.

CFR (Code of Federal Regulations). Title 49 Parts 171-180. Hazardous Materials Regulations. Department of Transportation. Research and Special Programs Administration. Docket HM-181. Final Rule 55 FR 52402 December 21, 1990; Revisions 56 FR 66124 December 20, 1991; revisions (to 49 CFR 171) 58 FR 12181.

Chang-Yeung, M., and J-L. Malo. 1994. Aetiological agents in occupational asthma. Eur. Respir. J. 7:346-71.

Chiodini, R. J., and J. P. Sundberg. 1981. Salmonellosis in reptiles: A review. Am. J. Epidemiol. 113: 494-9.

Chomel, B.B., R.W. Kasten, K. Floyd-Hawkins, B. Chi., K. Yamamoto, J. Roberts-Wilson, A.N. Gurfield, R.C. Abbott, N.C. Pederson, J.E. Koehler. 1996. Experimental transmission of Bartonella henselae by the cat flea. J. Clin. Microbiol. 34(8):1952-6.

Clark, A. J., P. M. Clissold, R. A. Shawi, P. Beatie, and J. Bishop. 1984. Structure of the mouse major urinary protein genes: different splicing configurations in the 3" -non-coding region. The EMBO J 3:1045-52.

Dalgard, D. W., R. J. Hardy, S. L. Pearson, G. J. Pucak, R. V. Quander, P. M. Zack, C. J. Peters, and P. B. Jahrling. 1992. Combined simian hemorrhagic fever and Ebola virus infection in cynomolgus monkeys. Lab. Anim. Sci. 42(2):152-7.

deBlay, F., M. D. Chapman, and T. A. E. Platts-Mills. 1991. Airborne cat allergen (*Fel d* 1): environmental control with the cat in situ. Am. Rev. Respir. Dis. 143:1334-9.

deGroot, H., K. G. H. Goei, P. VanSwieten, and R. C. Aalberse. 1991. Affinity purification of a major and a minor allergen from dog extract: serologic activity of affinity-purified *Can f* 1 and of *Can f* 1-depleted extract. J. Allergy Clin. Immunol. 87:1056-65.

Deming, M. S., R. V. Tauxe, and P. A. Blake. 1987. *Campylobacter enteritis* at a university: transmission from eating chicken and from cats. Am. J. Epidemiol. 126:526.

Des Prez, R. M., and C. R. Heim. 1990. *Mycobacterial tuberculosis*. Pp. 1877-1906 in Principles and Practices of Infectious Diseases, G. L. Mandell, D. R. Gordon, and J. E. Bennett, eds. New York: Churchill Livingstone.

Desjardins, A., C. Benoit, H. Chezzo, J. L'Archevêque, C. Leblanc, L. Paquette, A. Cartier, and J. L. Malo. 1993. Exposure to domestic animals and risk of immunologic sensitization in subjects with asthma. J. Allergy Clin. Immunol. 91:979-86.

DiBerardinis, L. J., J. S. Baum, M. W. First, G. T. Gatwood, E. F. Groden, and A. K. Seth. 1993. Guidelines for Laboratory Design: Health and Safety Considerations. 2nd ed. New York: John Wiley & Sons. 514 pp.

Dreeszen, P. H. 1993. Milwaukee illness: a sick municipal water system's potential threat to lab animals. Lab Anim. 22:36-40.

Dubey, J. P., and J. L. Carpenter. 1993. Histologically confirmed clinical toxoplasmosis in cats: 100 cases (1952-1990). J. Am. Vet. Med. Assoc. 203:1556-66.

Dykewicz, C. A., V. M. Dato, S. P. Fisher-Hoch, M. V. Horwath, G. I. Perez-Oronoz, S. M. Ostroff, H. Gary, Jr., L. B. Schonberger, and J. B. McCormick. 1992. Lymphocytic choriomeningitis outbreak associated with nude mice in a research institute. J. Am. Med. Assoc. 267:1349-53.

Edwards, R. G., M. F. Beeson, and J. M. Dewdney. 1983. Laboratory animal allergy: the measurement of airborne urinary allergens and effects of different environmental conditions. Lab. Anim. 17:235-9.

Eggleston, P. A., C. A. Newill, A. A. Ansari, A. Pustelnik, S-R. Lou, R. Evans III, D. G. Marrh, J. L. Longbottom, and M. Corn. 1989. Task-related variation in airborne concentrations of laboratory animal allergens: studies with *Rat n* I. J. Allergy Clin. Immunol. 84:347-52.

Eggleston, P. A., A. A. Ansari, B. Ziemann, N. F. Adkinson, Jr., and M. Corn. 1990. Occupational challenge studies with laboratory workers allergic to rats. J. Allergy Clin. Immunol. 86:63-72.

Erickson, G. A., E. A. Cabrey and G. A. Gustafson. 1975. Generalized contagious ecthyma in a sheep rancher: diagnostic consideration. J. Am. Vet. Med. Assoc. 166:262-3.

Fayer, R., and B. L. P. Ungar. 1986. Cryptosporidium spp. and cryptosporidiosis. Microbiol. Rev. 50:458-83.

Findlay, S. R., E. Stosky, K. Leitermann, Z. Hemady, and J. L. Ohman, Jr. 1983. Allergens detected in association with airborne particles capable of penetrating into the peripheral lung. Am. Rev. Respir. Dis. 128:1008-12.

Fjeldsgaard, B. E., and B. Smestad Paulsen. 1993. Comparison of IgE-binding antigens in horse dander and a mixture of horsehair and skin scrapings. Allergy 48:535-41.

Fleming, D. O., J. Richardson, J. Tulis, and D. Vesley, eds. 1995. Laboratory Safety: Principles and Practices. 2nd edition. Washington, D.C.: ASM Press. 406 pp.

Flynn, R. J. 1973. Parasites of Laboratory Animals. Ames: Iowa State University Press.

Fowler, M. E. 1986. Restraint. Pp. 38-50 in Zoo and Wildlife Animal Medicine, 2nd Edition, M. E. Fowler, ed. Philadelphia: W. B. Saunders.

Fox, J. G. 1982. Campylobacter: a "new" disease in laboratory animals. Lab. Anim. Sci. 32:625.

Fox, J. G., C. E. Newcomer, and H. Rozmiarek. 1984. Selected zoonoses and other health hazards. Pp. 614-48 in Laboratory Animal Medicine, J. G. Fox, B. J. Cohen and F. M. Loew, eds. Orlando: Academic Press.

Fox, J. G., N. S. Taylor, and J. L. Penner. 1989a. Investigation of zoonotic acquired *Campylobacter jejuni* enteritis with serotyping and restriction endonuclease DNA analysis. J. Clin. Micro. 27:2423.

Fox, J. G., K. O. Maxwell, and N. S. Taylor. 1989b. *Campylobacter upsaliensis* isolated from cats as identified by DNA relatedness and biochemical features. J. Clin. Microbiol. 27:2376.

Fox, J. G., and N. S. Lipman. 1991. Infections transmitted by large and small laboratory animals. Pp. 131-63 in Infectious Disease Clinics of North America, Vol. 5, R. C. Moellering, Jr., D. J. Weber, and A. N. Weinberg, eds. Philadelphia: W. B. Saunders.

Gajdusek, D. C. 1982. Muroidvirus nephropathies and muroid viruses of the hantaan virus group. Scand. J. Infect. Dis., Suppl. 36:96-108.

Gay, F. P., and M. Holden. 1933. The herpes encephalitis problem. II. J. Infect. Dis. 53:287-303.

Gnann, J. W., Jr., G. S. Bressler, C. A. Bodet, and C. K. Avent. 1983. Human blastomycosis after a dog bite. Ann. Int. Med. 98:48-9.

Geller, E. H. 1979. Health hazards for man. Pp. 402-7 in The Laboratory Rat, H. J. Baker, J. R. Lindsey, and S. H. Weisbroth, eds. New York: Academic Press.

Goldstein, E. J. C. 1990a. Bites. Pp. 834-7 in Principles and Practices of Infectious Diseases, G. L. Mandell, D. R. Gordon, and J. E. Bennett, eds. New York: Churchill Livingstone.

Goldstein, E. J. C. 1990b. Household pets and human infections. Pp. 117-130 in Infectious Disease Clinics of North America, Vol. 5, R. C. Moellering, Jr., D. J. Weber, and A. N. Weinberg, eds. Philadelphia: W. B. Saunders.

Gordon, S., R. D. Tee, D. Lowson, J. Wallace, and A. J. Newman-Taylor. 1992. Reduction of airborne allergenic urinary problems from laboratory rats. Br. J. Ind. Med. 49:416-22.

Grandin, T. 1987. Animal handling. Vet. Clin. N. Am.—Large Animal. 3(2):323-38.

Grist, N. R., and J. A. N. Emslie. 1985. Infections in British clinical laboratories, 1982-3. J. Clin. Pathol. 38:721-5.

Grist, N. R., and J. A. N. Emslie. 1987. Infections in British clinical laboratories, 1984-5. J. Clin. Pathol. 40:826-9.

Groves, M. G., J. D. Hoskins, and K. S. Harrington. 1993. Cat scratch disease: An update. Compendium 15:441-8.

Gutman, A. A., and R. K. Bush. 1993. Allergens and other factors important in atopic disease. Pp. 93-158 in Allergic Diseases: Diagnosis and Management. R. Patterson, L. C. Grammer, P. A. Greenberger, and C. R. Zeiss, eds. Philadelphia: J. B. Lippincott.

Hadley, K. M., D. Carrington, C. E. Crew, A. A. Gibson, and W. S. Hislop. 1992. Ovine chlamydiosis in an abattoir worker. J. Inf. 25 Suppl. 1:105-9.

Halstead, B. W. 1978. Poisonous and Venomous Marine Animals of the World. Princeton: Darwin Press.

Hanel, E., Jr., and R. H. Kruse. 1967. Laboratory-acquired mycoses. Department of the Army, Miscellaneous Publication 28.

Hanson, R. P., and others. 1950. Human infections with the virus of vesicular stomatitis. J. Lab. Clin. Med. 36:754-758.

Hanson, R. P., S. E. Sulkin, E. L. Buescher, W. McD. Hammond, R. W. McKinney, and T. E. Work. 1967. Arbovirus infections of laboratory workers. Science 158:1283-1286.

Hanson, L. E. 1982. Leptospirosis in domestic animals: the public health perspective. Am. J. Vet. Med. Assoc. 181:1505.

Harries, M. G., and O. Cromwell. 1982. Occupational asthma caused by allergy to pigs' urine. Br. Med. J. 284:867.

Heidam, L. Z. 1984. Spontaneous abortions among laboratory workers: A follow-up study. J. Epi. Comm. Health 38:36-41.

Hilliard, J. 1992. Diagnosis of *Herpesvirus simiae* infection. Res. Resources Reporter. Nat. Inst. Health. March. Washington, D.C.:NCRR.

Hollander, A., G. Doekes, and D. Heederik. 1996. Cat and dog allergy and total IgE as risk factors for laboratory animal allergy. J. Allergy Clin. Immunol. 98:545-54.

Hollinger, R. B., and P. A. Glombicki. 1990. Hepatitis A virus. Pp. 1383-95 in Principles and Practices of Infectious Diseases, G. L. Mandell, D. R. Gordon, and J. E. Bennett, eds. New York: Churchill Livingstone.

Holmes, G. P., J. K. Hilliard, K. C. Klontz, A. H. Rupert, C. M. Schindler, E. Parrish, D. G. Griffin, G. S. Ward, N. D. Bernstein, T. W. Bean, M. R. Ball, J. A. Brady, M. H. Wilder, and J. E. Kaplan. 1990. B birus (*Herpesvirus simiae*) infection in humans: epidemiologic investigation of a cluster. Ann. Intern. Med. 112:833-9.

Holmes, G. P., L. E. Chapman, J. A. Stewart, S. E. Straus, J. K. Hilliard, D. S. Davenport, and the B Virus Working Group. 1995. Guidelines for the prevention and treatment of B-virus infections in exposed persons. Clin. Inf. Dis. 20:421-39.

Hook, E. W. 1990. Salmonella species (including typhoid fever). Pp. 1700-13 in Principles and Practices of Infectious Diseases, G. L. Mandell, D. R. Gordon, and J. E. Bennett, eds. New York: Churchill Livingstone.

Hotchin, J., E. Sikora, W. Kinch, A. Hinman, and J. Woodal. 1974. Lymphocytic choriomeningitis in a hamster colony causes infection of hospital personnel. Science 185:1173-4.

Hunskaar, S., and R. Fosse. 1993. Allergy to laboratory mice and rats: a review of its prevention, management, and treatment. Lab. Anim. 27:206-21.

Hunt, R. D., W. W. Carlton, and N. W. King. 1978. Viral diseases. Pp. 1313 in Pathology of Laboratory Animals, K. Benischrke, F. M. Garner, and T. C. Jones, eds. New York: Springer-Verlag.

Huwyler, T., and B. Wüthrich. 1992. A case of fallow deer allergy. Allergy 47:574-5.

IPAC (Immunization Practices Advisory Committee). 1996. CDC (Center for Disease Control and Prevention) publishes frequent updates of recommended immunization practices. Contact CDC or visit their web site for specific agents and diseases of interest.

Jacobson, J. T., R. B. Orlob, and J. L. Clayton. 1985. Infections acquired in clinical laboratories in Utah. J. Clin. Microbiol. 21:486-9.

Jahrling, P. B. 1989. Arenaviruses and filoviruses. Pp. 857-891 in Diagnostic Procedures for Viral, Rickettsial and Chlamydial Infections, N. J. Schmidt and R. W. Emmons, eds. Washington, D. C.: American Public Health Association.

Jahrling, P. B., and C. J. Peters. 1992. Lymphocytic choriomeningitis virus, a neglected pathogen of man. Arch. Pathol. Lab. Med. 116:486-8.

Jeanselme, E., and P. Chevallier. 1910. Chancres sporotrichosiques des doigts produits par la morsur d'un rat inocule de sporotrichose. Bull. Mem. Soc. Med. Hop. (Paris) 30:176-8.

Jeanselme, E., and P. Chevallier. 1911. Transmission de la sporotrichose a l'homme par les morsures d'un rat blanc inocule avec une nouvelle variete de *Sporotrichum*: lymphangite gommeuse ascendante. Bull. Mem. Soc. Med. Hop. (Paris) 31:287-301.

Johnson, K. M. 1990a. Marburg and ebola viruses. Pp. 1303-6 in Principles and Practices of Infectious Diseases, G. L. Mandell, D. R. Gordon, and J. E. Bennett, eds. New York: Churchill Livingstone.

Johnson, K. M. 1990b. Lymphocytic choriomeningitis virus, lassa virus (Lassa Fever) and other arenaviruses. Pp. 1329-34 in Principles and Practices of Infectious Diseases, G. L. Mandell, D. R. Gordon, and J. E. Bennett, eds. New York: Churchill Livingstone.

Jordan, H. E., S. T. Mullins, and M. E. Stebbins. 1993. Endoparasitism in dogs: 21,583 cases (1981-1990). J. Am. Vet. Med. Assoc. 203:547-9.

Katz, D., J. K. Hilliard, R. Eberle, and S. L. Lipper. 1986. ELISA for detection of group common and virus specific antibodies in human and simian sera induced by herpes simplex and related simian viruses. J. Virol. Methods 14:99-109.

Kaufman, A. F., J. L. Moulthrop, and R. M. Moore. 1972. A perspective of simian tuberculosis in the United States-1972. J. Med. Primatol. 4:278-86.

Kaufmann, A. F., and D. C. Anderson. 1978. Tuberculosis control in nonhuman primates. Pp. 227-34 in Mycobacterial Infections of Zoo Animals, R. J. Montali, ed. Washington, D.C.: Smithsonian Institution Press.

Kawamata, J., T. Yamanouchi, K. Dohmae, H. Miyamoto, M. Takahaski, K. Yamanishi, T. Kurata, and H. W. Lee. 1987. Control of Laboratory acquired hemorrhagic fever with renal syndrome (HFRS) in Japan. Lab. Anim. Sci. 37:431-6.

Keimig, S. N., N. Esmen, S. Erdal , and E. Sansone. 1991. Solvent desorption from carbon beds in ducted and non-ducted laboratory fume hoods. Appl. Occupational Environ. Hyg. 67(7):592-597.

REFERENCES

Kesel, M. L. 1990. Handling, restraint, and common sampling and administration techniques in laboratory species. Pp. 333-61 in The Experimental Animal in Biomedical Research, Vol. 1, B. E. Rollin and M. L. Kesel, eds. Boca Raton: CRC Press.

Khabbaz, R. F., T. Rowe, M. Murphey-Corb, W. M. Heneine, C. A. Schable, J. R. George, C. P. Pau, B. S. Parekh, M. D. Lairmore, J. W. Curran, J. E. Kaplan, G. Schochetman, and T. M. Folks. 1992. Simian immunodeficiency virus needlestick accident in a laboratory worker. Lancet 340:271-3.

Kirkpatrick, C. E. 1990. Enteric protozoal infections. Pp. 804-14 in Infectious Diseases of the Dog and Cat, C. E. Greene, ed. Philadelphia: W. B. Saunders.

Kleine-Tebbe, J., A. Kleine-Tebbe, S. Jeep, C. Schou, H. Løwenstein, and G. Kunkel. 1993. Role of the major allergen (Fel d 1) in patients sensitized to cat allergens. Int. Arch. Allergy Immunol. 100:256-62.

Knysak, D. 1989. Animal aeroallergens. Immunol. Allergy Clin. No. Amer. 9:357-64.

Koehler, J. E., C. A. Glaser, and J. W. Tappero. 1994. Rochalimaea henselae infection: a new zoonosis with the domestic cat as a reservoir. J. Am. Med. Assoc. 271:531-5.

Kornegay, R. W., W. E. Giddens, Jr., G. L. Van Hoosier, Jr., and W. R. Morton. 1985. Subacute nonsuppurative hepatitis associated with hepatitis B virus infection in two cynomolgus monkeys. Lab. Anim. Sci. 35:400-4.

Kruse, R. M., W. M. Puckell, and J. M. Richardson. 1991. Biological Safety Cabinetry. Clin. Microbiol Rev. 8:207-241.

Ladiges, W. C., R. F. DiGiacomo, and R. A. Yamaguchi. 1982. Prevalence of Toxoplasma gondii antibodies and oocysts in pound-source cats. J. Am. Vet. Med. Assoc. 180:1334-5.

Lairmore, M. D., J. E. Kaplan, M. D. Daniel, N. W. Lerche, P. L. Nara, H. M. McClure, J. W. McVicar, R. W. McKinney, M. Hendry, P. Gerone, M. Rayfield, D. O. Johnsen, R. Purcell, J. Gibbs, J. Allan, J. L. Ribas, H. J. Klein, P. B. Jahrling, and B. Brown. 1989. Guidelines for the prevention of simian immunodeficiency virus infection in laboratory workers and animal handlers. J. Med. Primatol. 18:167-74.

Langley, J. M., T. H. Marrie, and A. Covert. 1988. Poker player's pneumonia: an urban outbreak of Q fever following exposure to a parturient cat. N. Engl. J. Med. 319:354-6.

Lankas, G. R., and R. D. Jensen. 1987. Evidence of hepatitis A infection in immature rhesus monkeys. Vet. Pathol. 24:340.

Laperche, Y., K. R. Lynch, K. P. Dolan, P. Piegelson. 1983. Tissue-specific control of $\alpha_{2\mu}$glubulin gene expression: constitutive syntheses in the submaxillary gland. Cell 32:453-60.

Lee, H. W., and K. M. Johnson. 1982. Laboratory-acquired infections with hantaan virus, the etiologic agent of Korean hemorrhagic fever. J. Inf. Dis. 146:645-51.

Legge, M. 1986. Reproductive hazards in laboratory environments. Australian J. Med. Lab. Sci. 7:44-7.

Le Guenno, B. P., P. Formentry, M. Wyers, P. Gounon, F. Walker, C. Boesch. 1995. Isolation and partial characterization of a new strain of Ebola virus. Lancet 345:1271-1274.

Longbottom, J. L. 1980. Purification and characterization of allergens from the urines of mice and rats. P. 483-90 in Oehling, A. I. Glazer, E. Mathov, and C. Arbesman, eds. Advances in Allergology and Appl Immunol. Oxford: Pergamon Press.

Lorusso, J. R., S. Moffat, and J. L. Ohman, Jr. 1986. Immunologic and biochemical properties of the major mouse urinary allergen (Mus m 1). J. Allergy Clin. Immunol. 78:928-37.

Luczynska, C. M., Y. Li, M. D. Chapman, and T. A. E. Platts-Mills. 1990. Airborne concentrations and particle size distribution of allergen derived from domestic cats (Felis domesticus). Am. Rev. Respir. Dis. 141:361-7.

LeDuc, J. W. 1987. Epidemiology of hantaan and related viruses. Lab. Anim. Sci. 37:413-8.

Lutsky, I. 1987. A worldwide survey of management practices inn laboratory animal allergy. Ann. Allergy 58:243-7.

Malo, J-L., and A. Cartier. 1993. Occupational reactions in the seafood industry. Chin. Reviews Allergy 11:223-40.

Marini, R. P., J. A. Adkins, and J. G. Fox. 1989. Proven or potential zoonotic diseases of ferrets. Am. J. Vet. Med. Assoc. 195:990.

Marrie, T. J., W. F. Schlech, C. J. Williams, and L. Yates. 1990. Q fever pneumonia associated with exposure to wild rabbits. Lancet 1:427-9.

Marsh, D. G., J. D. Neely, D. R. Breazeale, B. Ghosh, L. R. Friedhoff, E. Ehrlich-Kautzky, C. Schou, G. Kwishnaswamy, and T. H. Beaty. 1994. Linkage analysis of IL-4 and other chormosome 5Q31.1 markers and total immunoglobulin E concentrations. Science 264:1252-5.

Martini, G. A., and R. Siegert, eds. 1971. Marburg virus disease. Virology Monograph, Vol. 11. Vienna: Springer-Verlag.

Martini, G. A. 1973. Marburg virus. Postgrad. Med. J. 49:542.

Matson, S. C., M. C. Swanson, C. E. Reed, and J. W. Yunginger. 1983. IgE and IgG-immune mechanisms do not mediate occupation-related respiratory or systemic symptoms in hog farmers. J. Allergy Clin. Immunol. 72:299-304.

McAleer, R. 1980. An epizootic in laboratory guinea pigs due to *Trichophyton mentagrophytes*. Aust. Vet. J. 56:234-6.

McNulty, W. P. 1968. A pox disease of monkeys transmitted to man: clinical and histological features. Arch. Dermatol. 97:286-93.

McGrail, M.P., S.P. Tsai and E.J. Bernacki. 1995. A comprehensive initiative to manage the incidence and cost of occupational injury and illness. J. Occup. Environ. Med. 37(11):1263-1268.

Middleton, E., Jr. 1991. Asthma, inhaled allergens, and washing the cat. Am. Rev. Respir. Dis. 143:1209-10.

Miller, C. D., J. R. Songer, and J. F. Sullivan. 1987. A twenty-five year review of laboratory acquired human infections at the National Animal Disease Center. Am. Ind. Hyg. Assoc. J. 48:271-5.

Miller, R. A., M. A. Bronsdon, L. Kuller, and W. R. Morton. 1990. Clinical and parasitological aspects of cryptosporidiosis in nonhuman primates. Lab. Anim. Sci. 40:42-6.

Moore, R. M., Jr., B. R. Zehmer, J. I. Moultrop, and J. L. Parker. 1977. Surveillance of animal-bite cases in the United States. Arch. Environ. Health 32:267-70.

Morgenstern, J. P., I. J. Griffith, A. W. Brauer, B. L. Rogers, J. F. Bond, M. D. Chapman, and M. C. Kuo. 1991. Amino acid sequence of *Fel d* 1, the major allergen of the domestic cat: protein sequence analysis and cDNA cloning. Proc. Natl. Acad. Sci. 8:9690-4.

Morrison, Y. Y., and R. C. Rathbun. 1995. Hantavirus pulmonary syndrome: the Four Corners disease. Ann. Pharmacotherapy 29:57-65.

Mumford, R. S., R. D. Weaver, C. Patton, J. C. Feely, and R. H. Feldman. 1975. Canine brucellosis: a clinical and epidemiological study of two cases. J. Am. Med. Assoc. 321:1267-9.

Munoz, R. M., S. L. Lipper, and J. K. Hilliard. 1988. Identification of *Herpesvirus simiae* type specific polypeptides in a human outbreak of this virus. Abstract 198 in 13th Annual Herpesvirus Workshop. Irvine: University of California.

Murphy, F. A., M. P. Kiley, and S. P. Fisher-Hoch. 1990. Filoviridae. Pp. 933-42 in Virology, 2nd edition., B. N. Fields and D. M. Kmpe, eds. New York: Raven Press.

Nenzen, B. 1990. Cancer: a threat to laboratory personnel. Arbetsmiljoe 3:10-2.

Newill, C. A., V. L. Prenger, J. E. Fish, R. Evans III, E. L. Diamond, Q. Wei, and P. A. Eggleston. 1992. Risk factors for increased airway responsiveness to methacholine challenge among laboratory animal workers. Am. Rev. Respir. Dis. 146:1494-1500.

Nicklas, W. 1987. Introduction of salmonellae into a centralized laboratory animal facility by infected day old chicks. Lab. Anim. 21:161-3.

Nicklas, W., V. Kraft, and B. Meyer. 1993. Contamination of transplantable tumors, cell lines, and monoclonal antibodies with rodent viruses. Lab. Anim. Sci. 43:296-300.

NIH (National Institutes of Health). 1994. NIH Guidelines for Research Involving Recombinant DNA Molecules. National Institutes of Health. 59 FR 34496 June 24, 1994. Amended 59 FR 40170, August 5, 1994. Amended 60 FR 20726, April 27, 1995. Amended 61 FR 1482, January 19, 1995. Amended 61 FR 10004, March 12, 1996.

NIOSH (National Institute for Occupational Safety and Health). 1991. Work Practices Guide for Manual Lifting. DHHS Pub. No. 81-122. Washington, D.C.: US Government Printing Office.

NRC (National Research Council), Institute of Laboratory Animal Resources Subcommittee on Care and Use of Nonhuman Primates. 1980. Laboratory Animal Management: Nonhuman Primates. Washington, D.C.: National Academy Press.

NRC (National Research Council), Committee on Hazardous Substances in the Laboratory. 1981. Prudent Practices for Handling Hazardous Chemicals in Laboratories. Washington, D.C.: National Academy Press.

NRC (National Research Council), Committee on Hazardous Biological Substances in the Laboratory. 1989. Biosafety in the Laboratory: Prudent Practices for the Handling and Disposal of Infectious Materials. Washington, D.C.: National Academy Press.

NRC (National Research Council), Committee on the Study of Prudent Practices for Handling, Storage, and Disposal of Chemicals in Laboratories. 1995. Prudent Practices in the Laboratory: Handling and Disposal of Chemicals. Washington, D.C.: National Academy Press.

NRC (National Research Council), Institute of Laboratory Animal Resources Committee to Revise the Guide for the Care and Use of Laboratory Animals. 1996. Guide for the Care and Use of Laboratory Animals, 7th edition.Washington, D.C.: National Academy Press.

NSC (National Safety Council). 1988. Fundamentals of Industrial Hygiene, 3rd edition. B.A. Plog, ed. Chicago: NSC.

NSF (National Sanitation Foundation). 1992. Standard 49, Class II (Laminar Flow) Biohazard Cabinetry. Ann Arbor, MI.

Ohman, J. L., F. C. Lowell, and K. J. Bloch. 1974. Allergens of mammalian origin III. Properties of a major feline allergen. J. Immunol. 13:1668-77.

Ohman, J. H., H. Baer, M. C. Anderson, K. Leiterman, and P. Brown. 1983. Surface washing of living cats: an improved method of obtaining clinically relevant allergen. J. Allergy Clin. Immunol. 72:288-93.

Olfert, E. 1993. Allergies to laboratory animals: aspects of monitoring and control. Lab. Anim. Feb 1993:32-5.

Palmer, A. E. 1987. *Herpesvirus simiae*: historical perspective. J. Med. Primatol. 16:99-130.

Patterson, W. C., L. O. Mott, and E. W. Jenney. 1958. A study of vesicular stomatitis in man. J. Am. Vet. Med. Assoc. 133(1):57-62.

Petry, R. W., M. J. Voss, L. A. Kroutil, W. Crowley, R. K. Bush, and W. W. Busse. 1985. Monkey dander asthma. J. Allergy Clin. Immunol. 75:268-71.

Pike, R. M. 1976. Laboratory-associated infections: summary and analysis of 3,921 cases. Health Lab. Sci. 13:105-14.

Pike, R. M. 1979. Laboratory-associated infections: incidence, fatalities, causes and prevention. Ann. Rev. Microbiol. 33:41-66.

Platts-Mills, T. A. E., P. W. Heymann, J. L. Longbottom, and S. R. Wilkins. 1986. Airborne allergens associated with asthma: particle sizes carrying dust mite and rat allergens measured with a cascade impactor. J. Allergy Clin. Immunol. 77:850-7.

Price, J. A., and J. L. Longbottom. 1988. ELISA method for measurement of airborne levels of major laboratory animal allergens. Clin. Allergy 18:95-107.

Reese, N. C., W. L. Current, J. V. Ernst, and W. S. Barley. 1982. Cryptosporidiosis of man and calf: a case report and results of experimental infections in mice and rats. Am. J. Trop. Med. Hyg. 31:226-9.

Reijula, K., T. Virtanen, L. Halmepuro, H. Anttonen, R. Mäntyjäri, and J. Hassi. 1992. Detection of airborne reindeer epithelial antigen by enzyme-linked immunosorbent assay inhibition. Allergy 47:203-6.

Richardson, J. H. 1973. Provisional summary of 109 laboratory-associated infections at the Centers for Disease Control, 1947-1973. Presented at the 16th Annual Biosafety Conference, Ames, Iowa. [Abstract]

Richter, C. B., N. D. M. Lehner, and R. V. Henrickson. 1984. Primates. Pp. 298-373 in Laboratory Animal Medicine, J. G. Fox, B. J. Cohen, and F. M. Loew, eds. Orlando: Academic Press.

Rollag, O. J., M. R. Skeels, L. J. Nims, J. P. Thilsted, and J. M. Mann. 1981. Feline plague in New Mexico: a report of five cases. J. Am. Vet. Med. Assoc. 179:1381-3.

Rosner, W. W. 1987. Bubonic plague. J. Am. Vet. Med. Assoc. 191:406.

Russell, F. E. 1983. Snake Venom Poisoning. Great Neck: Scholium International.

Russell, R. G., J. I. Sarmiento, and J. G. Fox. 1990. Evidence of reinfection with multiple strains of *C. jejuni* and *C. coli* in *Macaca nemistrina* housed under hyperendemic conditions. Infect. Immun. 58:2149.

Ruys, T. Ed. 1991. Handbook of Facilities Planning. Vol. 2: Laboratory Animal Facilities. New York: Van Nostrand Reinhold. 422 pp.

Saah, A. J. 1990. Rickettsiosis. Pp. 1463-5 in Principles and Practices of Infectious Diseases, G. L. Mandell, D. R. Gordon, and J. E. Bennett, eds. New York: Churchill Livingstone.

Sakaguchi, M. S. Inouye, H. Miyazawa, H. Kamimura, M. Kimura, and S. Yamazai. 1989a. Particle size of airborne mouse crude and defined allergens. Lab. Anim. Sci. 39:234-6.

Sakaguchi, M., S. Inoye, H. Miyazawa, H. Kamimura, M. Kimura, and S. Yamazaki. 1989b. Evaluation of dust respirators for elimination of mouse aeroallergens. Lab. Anim. Sci. 39:63-6.

Sakaguchi, M., S. Inouye, H. Miyazawa, H. Kamimura, M. Kimura, and S. Yamazaki. 1990. Evaluation of countermeasures for reduction of mouse airborne allergens. Lab. Anim. Sci. 40:613-5.

SALS (Subcommittee on Arbovirus Laboratory Safety) 1980. Laboratory safety for arboviruses and certain other viruses of vertebrates. Am. J. Trop. Med. Hyg. 29(6):1359-81.

Sanford, J. P. 1985. Snake bites. Pp. 1841-3 in Cecil Textbook of Medicine, J. G. Wyngaarden and L. H. Smith, Jr., eds. Philadelphia: W. B. Saunders.

Saunders, Jr., W. E., and E. A. Horowitz. 1990. Other mycobacterial species. Pp. 1914-25 in Principles and Practices of Infectious Diseases, G. L. Mandell, D. R. Gordon, and J. E. Bennett, eds. New York: Churchill Livingstone.

Schachter, J., H. B. Oster, and K. B. Meyer. 1969. Human infection with the agent of feline pneumonitis. Lancet 1:1063-5.

Schachter, J., and C. R. Dawson. 1978. Human Chlamydial Infections. Littleton: PSG Publishing.

Schou, C., V. G. Svendsen, and H. Løwenstein. 1991a. Purification and characterization of the major dog allergen, *Can f* 1. Clin. Exp. Allergy 21:321-8.

Schou, C., G. N. Hansen, T. Lintner, and H. Løwenstein. 1991b. Assay for the major dog allergen, *Can f* 1: Investigation of house dust samples and commercial dog extracts. J. Allergy Clin. Immunol. 88:847-53.

SDS (Supplementary Data System). 1986. Washington, D.C.: US Bureau of Labor Statistics.

Shapiro, J. 1990. Radiation Protection: A Guide for Scientists and Physicians. 3rd ed. Cambridge, Mass.: Harvard University Press.

Simpson, D. I. H., E. D. T. Bowen, and W. F. Bright. 1968. Vervet monkey disease: experimental infection of monkeys with the causative agent and antibody studies in wild-caught monkeys. Lab. Anim. 2:75.

Slavin, R. G. 1993. Contact dermatitis. Pp. 553-8 in Allergic Diseases: Diagnosis and Management. R. Patterson, L. C. Grammer, P. A. Greenberger, and C. R. Zeiss, eds. Philadelphia: J. B. Lippencott.

Soave, R., and C. S. Weikel. 1990. Cryptosporidium and other protozoa including *Isospora, Sarcocystis, Balantidium coli* and *Blastocystis*. Pp. 2122-30 in Principles and Practices of Infectious Diseases, G. L. Mandell, D. R. Gordon, and J. E. Bennett, eds. New York: Churchill Livingstone.

Spinelli, J. S.,M. S. Ascher, D. L. Brooks, S. K. Dritz, H. A. Lewis, R.H. Morrish, L. Rose, and R. Ruppanner. 1981. Q fever crisis in San Francisco: controlling a sheep zoonosis in a lab animal facility. Lab. Anim. 10(3):24-7.

Spitzauer, S., C. Schwiger, J. Anrather, C. Ebner, O. Scheiner, D. Kraft, and H. Rumpold. 1993. Characterization of dog allergens by means of immunoblotting. Int. Arch. Allergy Immunol. 100:60-7.

Spitzauer, S., C. Schweiger, W. R. Sperr, B. Pandjaitan, P. Valent, S. Muhl, C. Ebner, O. Scheiner, D. Kraft, H. Rumpold, and R. Balenta. 1994. Molecular characterization of dog albumin as a cross-reactive allergen. J. Allergy Clin. Immunol. 93:614-27.

Storz, J. 1971. Chlamydia and Chlamydia-Induced Diseases. Springfield, Ill.: Charles C Thomas.

Strandberg, M., K. Sandback, O. Axelson, and O. Sundell. 1970. Spontaneous abortion among women in hospital laboratories. Lancet 1:384-5.

Strickoff, R. S., and D. B. Walters. 1990. Laboratory Health and Safety Handbook. New York: Wiley-Interscience.

Sullivan, J. F., J. R. Songer, and I. E. Estrem. 1978. Laboratory-acquired infections at the National Animal Disease Center, 1960-1976. Health Lab. Sci. 15(1):58-64.

Sutherland, I., and I. Lindgren. 1979. The protective effect of BCG vaccination as indicated by autopsy studies. Tubercle 60:225-31.

Swanson, M. C., M. K. Agarwal, J. W. Yunginger, and C. E. Reed. 1984. Guinea-pig derived allergens. Am. Rev. Respir. Dis. 129:844-9.

Swanson, M. C., A. R. Campbell, M. T. O'Hollaren, and C. E. Reed. 1990. Role of ventilation, air filtration, and allergen production rate in determining concentrations of rat allergens in the air of animal quarters. Am. Rev. Respir. Dis. 141:1578-81.

Taylor, A. F., T. G. Stephenson, and H. A. Giese. 1984. Rat bite fever in a college student. MMWR 33:318.

Teasdale, E. L., G. E. Davies, and A. Slovak. 1993. Anaphylaxis after bites by rodents. Br. Med. J. 286:1480.

Tee, R. D., S. Gordon, M. Nieuwenhuijen, P. Cullinan, D. Lowson, and A. Neuman Taylor. 1993. Exposure-response relationships in rat exposed workers. J. Allergy Clin. Immunol. 91:239. [abstract]

Tribe, G. W., and M. P. Fleming. 1983. Biphasic enteritis in imported cynomolgus (*Macaca fascicularis*) monkeys infected with *Shigella, Salmonella* and *Campylobacter*. Lab. Anim. 17:65-9.

Tsai, T. F. 1987. Hemorrhagic fever with renal syndrome: mode of transmission to humans. Lab. Anim. Sci. 37:428-30.

Tsai, T. F. 1991. Arboviral infections in the United States. Pp. 73-102 in Infectious Disease Clinics of North America, Vol. 5, R. C. Moellering, Jr., D. J. Weber, and A. N. Weinberg, eds. Philadelphia: W. B. Saunders.

Tsai, T. F., S. P. Bauer, D. R. Sasso, S. G. Whitfield, J. B. McCormick, T. C. Caraway, L. MacFarland, H. Bradford, and T. Kurata. 1985. Serological and virological evidence of a hantaan virus-related enzootic in the United States. J. Infect. Dis. 152:126-36.

Twiggs, J. T., M. K. Agarwal, M. J. E. Dahlberg, and J. W. Yunginger. 1982. Immunochemical measurement of airborne mouse allergen in a laboratory animal facility. J. Allergy Clin. Immunol. 69:522-6.

Tzipori, S. 1988. Cryptosporidiosis in perspective. Adv. Parasitol. 27:63-129.

US Congress. 1971. Atomic Energy Act of 1946 and Amendments. Joint Committee on Atomic Energy. Washington, D.C.: US Government Printing Office.

VanMetre, T. E., Jr., D. G. Marsh, N. F. Adkinson, Jr., J. E. Fish, A. Kagey-Sobotka, P. S. Norman, E. B. Radden, Jr., and G. L. Rosenberg. 1986. Dose of cat (Felis domesticus) allergen 1 (*Fel d I*) that induces asthma. J. Allergy Clin. Immunol. 78:62-75.

Venables, K. M., R. D. Tee, E. R. Hawkins, D. J. Gordon, C. J. Wale, T. H. Lam, N. M. Farrer, P. J. Baxter, and A. J. Newman Taylor. 1988. Laboratory animal allergy in a pharmaceutical company. Br. J. Ind. Med. 45:660-6.

Wahn, U., and R. P. Siriganian. 1980. Efficacy and specificity of immunotherapy with laboratory animal allergen extracts. J. Allergy Clin. Immunol. 65:413-21.

Walls, A. F., A. J. Newman-Taylor, and J. L. Longbottom. 1985. Allergy to guinea pigs: identification of specific allergens in guinea pig dust by crossed radioimmunoelectrophoresis and investigation of the possible origin. Clin. Allergy 15:535-46.

Warner, J. A., and J. L. Longbottom. 1991. Allergy to rabbits. Allergy 46:481-91.

Watt, A. D. and C. P. McSharry. 1996. Laboratory animal allergy: anaphylaxis from a needle injury. Occup. Environ. Med. 53:573-74.

Weigler, B. J., D. W. Hird, J. K. Hilliard, N. W. Lerche, J. A. Roberts, and L. M. Scott. 1993. Epidemiology of cercopithecine Herpesvirus 1 (B virus) infection and shedding in a large breeding cohort of rhesus macaques. J. Infect. Dis. 167:257-67.

Weigler, B.J., F. Scinicariello, and J. K. Hilliard. 1995. Risk of venereal B virus (cercopithecine herpesvirus 1) transmission in rhesus monkeys using molecular epidemiology. J. Infect. Dis. 171:1139-43.

Wedum, A.G., W. E. Barkley, A. Hellman. 1972. Handling of Infectious agents. J. Am. Vet. Med. Assoc. 161:1557-1567.

Wells, D. L., S. L. Lipper, and J. Hilliard. 1989. *Herpesvirus simiae* contamination of primary rhesus monkey kidney cell cultures: CDC recommendations to minimize risks to laboratory personnel. Diagn. Microbiol. Infect. Dis. 12:333-5.

Wentz, P. E., M. C. Swanson, and C. E. Reed. 1990. Variability of cat-allergen shedding. J. Allergy Clin. Immunol. 85:94-8.

Werner, A. H., and B. E. Werner. 1993. Feline sporotrichosis. Comp. Cont. Educ. Pract. Vet. 15:1189-97.

Wolf, R. H., S. V. Gibson, E. A. Watson, and G. B. Baskin. 1988. Multidrug chemotherapy of tuberculosis in rhesus monkeys. Lab. Anim. Sci. 38:25-33.

Wood, R. A., M. D. Chapman, N. F. Adkinson, Jr., and P. A. Eggleston. 1989. The effect of cat removal on allergen content in household-dust samples. J. Allergy Clin. Immunol. 83:730-4.

Wood, R. A., K. E. Mudd, and P. A. Eggleston. 1992. The distribution of cat and dust mite allergens on wall surfaces. J. Allergy Clin. Immunol. 89:126-30.

Wood, R. A., A. N. Laheri, and P. A. Eggleston. 1993. The aerodynamic characteristics of cat allergen. Clin. Exp. Allergy 23:733-9

Woodfolk, J., C. Luczynska, F. deBlay, M. D. Chapman, and T. A. E. Platts-Mills. 1992. Cat allergy. Ann. Allergy 69:273-5.

Xu, Z. Y., Y. W. Tang, L. Y. Kan, and T. F. Tsai. 1987. Cats—source of protection or infection? Am. J. Epidemiol. 126:942-948.

Ylönen, J., R. Mäntyjäri, A. Taivainen, and T. Virtanen. 1992. IgG and IgE antibody response to cow dander and urine in farmers with cow-induced asthma. Clin. Exp. Allergy 22:83-90.

Ylönen, J., T. Virtanen, L. Hormanheimo, S. Parkkinen, J. Pelkonen, and R. Mäntyjärvi. 1994. Affinity purification of the major bovine allergen by a novel monoclonal antibody. J. Allergy Clin. Immunol. 93:851-858.

Zhou, C., T. S. Hurst, D. W. Cockcroft, and J. A. Dorsman. 1991. Increased airway responsiveness in swine farmers. Chest 99:941-944.

Ziemann, B., M. Corn, A.A. Ansari, and P. Eggleston. 1992 The effectiveness of the Duo-flo Bioclean Unit for controlling airborne antigen levels. Am. Ind. Hyg. Assoc. J. 53:138-45.

Zwartouw, H. W., and E. A. Boulter. 1984. Excretion of B virus in monkeys and evidence of genital infection. Lab. Anim. 18:65-70.

Index

A

AAAAI. *See* American Academy of Allergy, Asthma, and Immunology
Abscesses, 34
Access control barriers, 108-109, 126
Accidents, 15, 25
Accountability and responsibility, 4, 15-17
Acoustical hazards. *See* Hazards, physical
Administration and management, 2, 4, 8, 16-18, 22, 121. *See also* Commitment, element of
consistency in, 4, 13, 15, 124
procedures, 6, 107
style and structure of, 23
Adverse-reaction report data, 27, 29
Aerosol transmission, 45, 48-49, 55-56, 73, 76, 85
African green monkeys, 46, 70, 78
Agricultural Research Service (ARS), 26
AIDS. *See* Susceptibility of employees
Airborne exposures. *See* Aerosol transmission
Air hoods. *See* Respirators
Airway hyperresponsiveness. *See* Asthma
Allergens. *See* Hazards
Allergic conjunctivitis, 53

Allergic rhinitis, 53-54, 59-60
Amebiasis, 98-99
American Academy of Allergy, Asthma, and Immunology (AAAAI), 64
American National Standards Institute (ANSI), 117
Amphibians, 60
Anaphylaxis, 52-53, 64
Anesthetic gases, 42-43
Animal and Plant Health Inspection Service (APHIS), 26
Animal biosafety levels, 45, 81
    Level 2 practices, 49, 72, 77, 80, 88, 90, 92, 94, 100-101
    Level 3 practices, 49, 68, 73, 83, 86, 91
    Level 4 practices, 48, 70
Animal care and use programs, 4, 18-20, 22, 112. *See also* Site inspections
    addressing occupational health in, 8
    progress in, 1
Animal-control measures, 79
Animals. *See also* individual species
    species-specific responses to, 12, 107
    wild, 13, 66-67, 79-81, 101-102 (*See also* Rodents, wild)

147

ANSI. *See* American National Standards Institute (ANSI)
APHIS. *See* Animal and Plant Health Inspection Service (APHIS)
Approvals required, 21, 39, 48
Arboviral infections, 45-46, 80-81
ARS. *See* Agricultural Research Service
Arthritis, septic, 34
Arthropod-borne viruses. *See* Arboviral infections
Arthropod infestations, 101-102, 104-105
Asthma, 51, 53-54, 57-63
Atopy, 54
Authorizations required. *See* Approvals required

B

Back injuries, 40
Bacterial pathogens, 46-47, 85-95
Balantidiasis, 99
Benign epidermal monkeypox, 73-74
Biological hazards. *See* Hazards
Biological safety cabinets, 109-111, 117
Biosafety in Microbiological and Biomedical Laboratories, 45, 48, 125
Biosafety in the Laboratory: Prudent Practices for the Handling and Disposal of Infectious Materials, 48
Biosafety levels. *See* Animal biosafety levels
Birds, 46, 59, 81-82, 85, 87-88, 93, 104
Bites and scratches. *See* Hazards, physical
Bloodborne pathogens. *See* Hazards, biological
Breathing, difficulty in, 53
Brucellosis, 34, 46, 90-91
Bubonic plague, 89-90
Bursitis, 40
B-virus infection, 27, 34, 46, 66-68

C

Cage cleaning. *See* Housekeeping practices

Cages, 62, 109-110
Campylobacteriosis, 46, 92
Cardiopulmonary resuscitation, 64
Carpal-tunnel syndrome, 40
Cats, 57-58, 71, 79, 81, 83-85, 95-97, 99-100, 102-104
Cat-scratch fever, 34, 83-84
Cattle, 46, 58-59, 85, 91
Cellulitis, 34
Centers for Disease Control and Prevention (CDC), 5, 25, 45, 66, 68, 72
Cercopithecine herpesvirus 1 (CHV1). *See* B-virus infection
Chemical fume hoods. *See* Respirators
Chemical inventory data, 25
Chemical restraint, 15, 33, 35, 68
Chemicals, hazardous. *See* Hazards, chemical
Chemoprophylaxis, 86, 88
Chlamydiosis. *See* Psittacosis
Choriomeningitis, lymphocytic. *See* Lymphocytic choriomeningitis
CHV1. *See* B-virus infection
Cleaning hazards. *See* Housekeeping practices
Collaboration. *See* Coordination within programs
Commitment, element of, 4, 8, 13, 15, 23-24, 121
Compliance Database data, 29-30
Compliance issues, 17, 23, 107. *See also* Approvals required; Regulatory requirements
Compressed-gas cylinders, 36
Computer links, 120
Confidentiality, 118
Conjunctivitis, allergic, 53
Consultants, using, 30-31, 64, 129, 132
Contagious ecthyma, 34, 74-75
Coordination within programs, 2-5, 8, 13, 15, 18, 21, 65
Coxiella burnetii. See Q fever
Coyotes, 46
Credibility issues, 15
Cryptosporidiosis, 97-98

Custodial service. *See* Housekeeping practices
Cylinders, compressed-gas, 36

D

Deer, 59
Dermatitis
  atopic, 54
  contact, 59
  contagious pustular (*See* Orf disease)
Dermatomycosis, 47, 99-100
Disinfectants, 42-43
Documentation, 118
Dogs, 46-47, 58, 79, 81, 84-85, 90-91, 95, 97, 99-100, 102-105
Dust-mist respirators. *See* Respirators

E

Ebola-Reston virus, 69
Ebola-virus infection, 68-70
Ecthyma, contagious. *See* Orf disease
Ectoparasitic infestations, 34, 74-75
Eczema. *See* Dermatitis, atopic
Edema, laryngeal, 53
Education and training, 4, 6, 12-14, 17, 34, 39-40, 62-63, 65, 114-116
Electric hazards. *See* Hazards, physical
Electromagnetic radiation. *See* Ionizing radiation
ELISA. *See* Enzyme-linked immunosorbent assay
Emergency procedures, 6, 120-121. *See also* Accidents
Employees
  defined (*See* Occupational health and safety programs, participation in)
  at risk, 4, 12, 17, 22, 111, 120, 129-130 (*See also* Hazards, perceived; Screening programs; Susceptibility of employees)
Endocarditis, 34
Engineering controls, 108-110
Enteric yersiniosis, 95

Envenomation, 34
Environmental health and safety programs, 2, 4, 6-7, 9, 18, 21-22, 29, 106
Environmental Protection Agency, 2
Enzyme-linked immunosorbent assay (ELISA), 61
Epinephrine, self-administered, 64
Episodic health evaluations, 10, 29, 131
Equipment, dangerous, 40-41
Equipment performance, 6, 25, 116-118
Ergonomic hazards, 40, 108. *See also* Hazards, physical
Evaluation
  program, 6, 121-122, 133-134
  of workers, 63-64, 124, 130-131, 133 (*See also* Episodic health evaluations)
Experimentation. *See* Research
Exposure, 4, 13-14, 43, 55, 125, 127
  controlling, 4, 25, 108-114

F

Facilities, 17, 107-108
  design, 6, 62
  operation, 6
Federal Bureau of Labor Statistics data, 27-28
Ferrets, 80, 85
Filoviruses. *See* Ebola-Reston
Filter-top cages, 62, 109-110
Fires. *See* Hazards, physical
First-aid log data, 27, 29, 134
First Report of Injury or Illness data, 27-28
Fish, 60
Flammable materials, classes of, 35
Flea-control measures, 84
Fleas. *See* Insects
Flight zone, 33
Fowl. *See* Birds
Fume hoods. *See* Respirators
Funding agencies, requirements of, 2
Fungal pathogens, 34, 47, 99-101

G

Gene therapy research, 48. *See also
Guidelines for Research Involving
Recombinant DNA Molecules*
Gerbils, 56
Giardiasis, 97
Goats, 46-47, 74-75, 85
*Guide for the Care and Use of Laboratory
Animals,* 63
Guidelines. *See* Rules and guidelines
*Guidelines for Research Involving
Recombinant DNA Molecules,* 48,
124
Guinea pigs, 46-47, 56, 95, 104

H

Hamsters, 46, 102
Hantavirus infection, 34, 71-72
Hazard Communication Standard, 42
Hazard information, sources of, 25-26
Hazards. *See also* Laboratory safety
   allergens, 3, 5-6, 24, 26, 51-64 (*See
also* Evaluation, of workers)
     mechanisms of, 53-54
     preventive measures, 60-64
   biological, 7, 127 (*See also* Approvals
required)
     infectious agents, 2, 7, 12-13, 44-48,
124
     radiation sources, 12, 20-21, 36-39
   chemical, 2, 5, 12-14, 24, 42-44
   defined, 26
   identifying (*See* Risk Assessment)
   novel, 47
   perceived, 24
   physical, 3, 5, 7, 24, 26, 32-41, 127
(*See also* Housekeeping practices)
     bites and scratches, 33-34, 44,
52-53
     electricity, 36
     fires, 35, 118
     noise, 41
     trauma, 40
   protocol-related (*See* Research,
protocol-related hazards of)

   undetected, 14, 31
   unknown, 44, 66
   zoonoses, 3, 6, 14, 20, 44-45, 65-105
HBV. *See* Hepatitis-B virus (HBV)
Health and safety risks. *See* Hazards
Health Research Extension Act of 1986,
124
Helminth-parasite infections, 101-103
Hemorrhagic fever. *See* Hantavirus
infection; Lymphocytic
choriomeningitis (LCM) virus
infection
HEPA filtration. *See* High-efficiency-
particulate-air-filtered (HEPA-
filtered) laminar-flow ventilation
Hepatitis A virus, 46, 76-77
Hepatitis B virus (HBV), 7, 77-78, 124
Hepatitis C, D, and E viruses, 77-78
High efficiency-particulate-air-filtered
(HEPA-filtered) laminar-flow
ventilation, 62, 110-112, 117
High-pressure lines, 36
Hives. *See* Anaphylaxis; Edema,
laryngeal; Uticaria, contact
Hoods. *See* Respirators
Horses, 59
Housekeeping practices, 39, 110, 112-113
Human-influenza viruses, 80
Human-resources function, 22
Husbandry, animal. *See* Animal care and
use programs
Hygiene, personal, 48-50, 63, 65, 111-112
Hypersensitivity, 54, 59

I

IACUC. *See* Institutional animal care and
use committee
ILAR. *See* Institute of Laboratory Animal
Resources
Immunizations. *See* Vaccinations
Immunotherapy, 64
Influenza-virus infections, 80
Information management, 6, 118-120. *See
also* Occupational health and safety
information, sources of

INDEX

Insects, 58, 60, 83-84
Inspections. *See* Site inspections
Institute of Laboratory Animal Resources, 25
Institutional animal care and use committee (IACUC), 2, 5, 18-21, 107, 121
Institutions
 categories of, 1-2
 functional parts of, 4, 16-22
 mission, 4, 123
 obligations of, 1, 16, 124
Interactions between departments, 18-19, 130, 134. *See also* Coordination within programs
Ionizing radiation, 38-39. *See also* Hazards, biological
IPAC. *See* Public Health Service Advisory Committee on Immunization Practices
Irradiation. *See* Ionizing radiation

J

Janitorial service. *See* Housekeeping practices
Job descriptions, 12, 22

L

Laboratory safety, 2, 42, 65, 85, 110-114
Laboratory Safety: Principles and Practices, 48
Laboratory standard. *See* Occupational Exposure to Hazardous Chemicals in Laboratories
Laryngeal edema, 53
Laser radiation, 37-38. *See also* Hazards, biological
Lassa fever, 72
LCM. *See* Lymphocytic choriomeningitis (LCM) virus infection
Lentivirus infections. *See* Simian immunodeficiency virus (SIV) infection
Leptospirosis, 47, 91-92

Lifting injuries, 40
Lighting, low, 36
Livestock, 46, 99. *See also* individual species
Lymphocytic choriomeningitis (LCM) virus infection, 14, 46, 65, 72-73

M

Macaques, 46, 69, 74, 78, 85, 125
Machinery, dangerous, 40-41
Management. *See* Administration and management
Marburg-virus disease, 46, 70
Material-Safety Data Sheets (MSDSs), 25, 35, 127
Measles, 75-76
Medical-surveillance programs, 78, 124, 134
Meningitis, 34
Mice, 46-47, 55-56, 91, 102, 104
Monitoring, 4, 13, 15, 22, 61, 116. *See also* Recordkeeping
 of effluent, 118
Monkeypox, 73-74
Monkeys, various, 76-78. *See also* African green monkeys; Macaques
Morbidity and Mortality Weekly Report, 66
MSDSs. *See* Material-Safety Data Sheets
Multidisciplinary approaches, 9

N

National Animal Disease Center (NADC), 26
National Center for Infectious Diseases, 72-73
National Fire Protection Association (NFPA), 35
National Institute for Occupational Safety and Health (NIOSH), 25, 40, 63
National Institutes of Health (NIH), 5, 26, 45, 68, 124-125
National Research Council, 1, 5
 recommendations by, 8-10

National Safety Council (NSC), 38
National Sanitation Foundation (NSF), 117
Near-miss reports, 131-132
Neuropathia endemica. *See* Hantavirus infection
Newcastle disease, 76
NFPA. *See* National Fire Protection Association
NIH. *See* National Institutes of Health
NIOSH. *See* National Institute for Occupational Safety and Health
Noise. *See* Hazards, physical
Nonparticulate radiation. *See* Ionizing radiation
NSC. *See* National Safety Council
NSF. *See* National Sanitation Foundation

O

Occupational Exposure to Hazardous Chemicals in Laboratories, 42
Occupational health and safety information, sources of, 26-30
Occupational health and safety programs, 18, 22
   costs of, 5, 30-31
   developing, 4-5, 17, 23-31
   elements of, 6, 106-122
   participation in, 9, 11-13, 123-124
Occupational health and safety risks. *See* Hazards
Occupational health-care service, 6-7, 123-134
   activities of, 129-133
   limitations of, 133
   responsibilities of, 125-129
Occupational health log data, 27, 29, 119
Occupational Safety and Health Administration (OSHA), 2, 7, 21, 25, 35, 41, 114, 121, 124
Occupations, categories of, 2
Orf disease, 34, 74-75
Orientation. *See* Education and training
Ornithosis. *See* Psittacosis
Orthopox virus. *See* Monkeypox

OSHA. *See* Occupational Safety and Health Administration (OSHA)
OSHA Form 101. *See* Supplementary Record of Occupational Injury or Illness (OSHA Form 101) data
OSHA 200 log data, 27-29, 119-120, 127, 134
Osteomyelitis, 34

P

Paramyxovirus infection. *See* Newcastle disease
Parrot fever. *See* Psittacosis
Personal protective equipment, 14, 17, 41, 49, 63, 111-114
Pesticides, 42-43
Physical examinations, 7, 9-10. *See also* Evaluation, of workers
Physical hazards. *See* Hazards
Pigs. *See* Swine
Plague, 89-90
Pneumonitis, 59
Poxvirus diseases. *See* Benign epidermal monkeypox; Monkeypox; Orf disease
Preexisting conditions, 60-61
Primates (nonhuman), 33, 46-47, 58, 76-77, 85-87, 94-95, 97, 99, 102-103
Priority-setting of hazards, 24
Professional societies, 25
Protective equipment. *See* Personal protective equipment
Protozoal pathogens, 95-99
Prudent Practices in the Laboratory: Handling and Disposal of Chemicals, 42, 44
Psittacosis, 46, 87-88
Public Health Service, 7
   Advisory Committee on Immunization Practices (IPAC), 132
Public Health Service Policy on Humane Care and Use of Laboratory Animals, 124
Pulmonary function, monitoring, 61, 64
Pustular dermatitis. *See* Orf disease

Q

Q fever, 14, 46, 65, 81-83

R

Rabbits, 47, 56, 85, 95, 99, 103-105
Rabies virus, 34, 79-80
Radiation, ionizing. *See* Ionizing radiation
Radioallergosorbent test (RAST), 61
Radioisotopes. *See* Hazards, biological
Ranking hazards, 26
RAST. *See* Radioallergosorbent test
Rat-bite fever, 34, 47, 88-89
Rats, 47, 54-55, 91, 102, 104
Recordkeeping, 4, 13, 15, 116, 132
Refresher training. *See* Education and training
Regulatory requirements, 2, 5, 7, 24, 112, 127
Reporting requirements, 15, 20, 131-132
Reptiles, 60, 93. *See also* Envenomation
Research, 4, 18, 21
  complexity of, 43-44
  protocol-related hazards of, 5, 7, 43-50
  reviews of protocols, 25, 115-116
Resource development. *See* Occupational health and safety programs, costs of
Respirators, 63, 109-111, 114
Responsibility and accountability, 4, 15-17
Restraint, 68, 113. *See also* Chemical restraint
Rhesus monkeys, wild-caught, 67
Rhinitis, allergic, 53-54, 59-60
Rickettsial pathogens, 46, 81-85
Risk
  assessment of, 4-9, 13-14, 20, 24-30, 125, 127
  defined, 26
Rodents, 99. *See also* Guinea pigs; Mice; Rats
  wild, 71-72, 84, 89-90, 104-105
Rubeola. *See* Measles
Rules and guidelines, 4, 13-15

S

Safety awards, 23
Safety bulletins and reports, 25
Safety cabinets, 109-111, 117
Salmonellosis, 47, 92-94
SALS. *See* Subcommittee on Arbovirus Laboratory Safety
Scratches. *See* Hazards, physical
Screening programs, 60-61, 126
SDS. *See* Supplementary Data System (SDS) data
Sedation. *See* Chemical restraint
Septic arthritis, 34
Serum collection, 7, 9-10, 126, 132-133
Sharps, controlling, 34-35, 48, 111
Sheep, 46, 59, 74-75, 81-82, 85
Shigellosis, 47, 94
Simian immunodeficiency virus (SIV) infection, 46, 78-79
Site inspections, 20-21, 25, 116, 134. *See also* Animal care and use programs
SIV. *See* Simian immunodeficiency virus (SIV) infection
Snake bite, 34
Sporotrichosis, 47, 100-101
Stomatitis virus, 46
Subcommittee on Arbovirus Laboratory Safety (SALS), 45, 70
Supplementary Data System (SDS) data, 27-28
Supplementary Record of Occupational Injury or Illness (OSHA Form 101) data, 28-29
Surveillance programs. *See* Medical-surveillance programs
Susceptibility of employees, 7, 127-128
  AIDS-related, 98
Swine, 46, 58, 85, 90-91, 97, 99, 103-104

T

Tanapox. *See* Benign epidermal monkeypox
Tennis elbow, 40
Tenosynovitis, 34
Tetanus, 34

Tissue-preserving chemicals, 42-43
Toxic chemicals, 42. *See also* Hazards, chemical
Toxoplasmosis, 95-97
Training. *See* Education and training
Transgenic animals, 48
Transplantation, 66
Tuberculosis, 47, 85-87
Tularemia, 34

U

Ultrasonography, hazards associated with, 41
Ultraviolet (UV) radiation. *See also* Hazards, biological
  biocidal use of, 117
  classification of, 36-37
Urticaria, contact, 52-53, 60
U.S. Army Medical Research Institute for Infectious Disease (USAMRIID), 83
U.S. Nuclear Regulatory Commission, 39
USAMRIID. *See* U.S. Army Medical Research Institute for Infectious Disease (USAMRIID)
UV. *See* Ultraviolet (UV) radiation

V

Vaccinations, 74, 76-80, 87, 90, 124, 126, 132
Ventilation systems, 62, 107-108, 117-118

Vesicular stomatitis virus, 46
Veterinarian, staff, 18, 20-21, 115, 127
Viral pathogens, 46, 66-81. *See also* individual viruses
Visibility, impaired, 36

W

Walk-through. *See* Site inspections
Welts. *See* Urticaria, contact
Wheezing. *See* Edema, laryngeal
Work environment. *See* Exposure, controlling
Worker compensation data, 26-28, 119, 127, 132, 134
Workplace diversity. *See* Occupational health and safety programs, participation in
Work practices. *See* Laboratory safety

X

Xenograft transplantation, 66

Y

Yersinia infection, 95

Z

Zoonoses. *See* Hazards